MODERN HUMANITIES RESEARCH ASSOCIATION
LIBRARY OF MEDIEVAL WELSH LITERATURE

GENERAL EDITORS
ANN PARRY OWEN
ERICH POPPE
SIMON RODWAY

Delw y Byd
A Medieval Welsh Encyclopedia

EDITED BY
NATALIA I. PETROVSKAIA

MODERN HUMANITIES RESEARCH ASSOCIATION
LIBRARY OF MEDIEVAL WELSH LITERATURE

ALREADY PUBLISHED

Welsh Court Poems
edited by Rhian M. Andrews (2007)
(available from University of Wales Press)

Selections from Ystorya Bown o Hamtwn
edited by Erich Poppe and Regine Reck (2009)
(available from University of Wales Press)

Early Welsh Gnomic and Nature Poetry
edited by Nicolas Jacobs (2012)
(available from http://www.mhra.org.uk/series/LMWL)

Historical Texts from Medieval Wales
edited by Patricia Williams (2012)
(available from http://www.mhra.org.uk/series/LMWL)

A Selection of Early Welsh Saga Poems
edited by Jenny Rowland (2014)
(available from http://www.mhra.org.uk/series/LMWL)

Arthur in Early Welsh Poetry
edited by Nerys Ann Jones (2019)
(available from http://www.mhra.org.uk/series/LMWL)

*Hystoria Gweryddon Yr Almaen: The Middle Welsh Life of
St Ursula And The 11,000 Virgins*
edited by Jane Cartwright (2020)
(available from http://www.mhra.org.uk/series/LMWL)

texts.mhra.org.uk

Delw y Byd

A Medieval Welsh Encyclopedia

Edited by
Natalia I. Petrovskaia

Modern Humanities Research Association
2020

Published by

The Modern Humanities Research Association
Salisbury House
Station Road
Cambridge CB1 2LA
United Kingdom

First published 2020

ISBN 978-1-78188-949-7 (HB)
ISBN 978-1-78188-950-3 (PB)

CONTENTS

LIST OF TABLES AND FIGURES

ACKNOWLEDGEMENTS

This edition, and the research leading up to it, were made possible through a Fellowship granted to me in 2015–2017 by the Alexander von Humboldt Foundation for a research stay at Philipps-Universität Marburg. My home institution, Utrecht University, generously allowed time free from teaching for the duration of this project.

The edition, and I myself, owe a profound debt of gratitude to the patience, expertise, and kindness of Professor Erich Poppe, my host during my Alexander von Humboldt Fellowship at Philipps-Universität Marburg, for his patience with the multiple drafts and versions of this text, and for his generosity with his time, spent debating minute details ranging from manuscript transmission to interpretation, to layout.

This project has its origins in my initial interest in medieval geographical perceptions during my PhD studies at the University of Cambridge, and I owe a profound debt of gratitude to Professor Paul Russell who first pointed me in the direction of this text, and who has patiently read and corrected various versions of my transcriptions, translations, and the edition itself.

Since I first commenced work on *Delw y Byd* around 2010 I have received the advice and support of many friends and colleagues at Cambridge, Marburg, and Utrecht, and at the various conferences where I presented elements of this research over the years. I am particularly grateful to Erich Poppe, Paul Russell, Peter Schrijver, Simon Rodway, Elena Parina, and Karianne Lemmen-Friedrich for reading and commenting on various drafts of this edition and of the accompanying glossary. All errors that might remain are, of course, my own responsibility.

I would also like to express my thanks to the MA Middle Welsh class of 2015–2016 at Philipps-Universität Marburg and the BA Middle Welsh 2 class of 2018–2019 at Utrecht University for their patience with and feedback on the edition and glossaries for individual chapters used during these classes. I also wish to thank Merel Reijntjens, student assistant in Celtic at Utrecht University (2018–2019), for her work in proofreading a late draft of the glossary. I also want to thank Gerard Lowe and everyone in the MHRA publications team who was involved with the production of this volume.

ABBREVIATIONS

BBCS *Bulletin of the Board of Celtic Studies*
BnF Bibliothèque nationale de France
CMCS *Cambridge/Cambrian Medieval Celtic Studies*
DB Lewis, Henry, and P. Diverres, eds, *Delw y Byd (Imago Mundi)* (Cardiff: University of Wales Press, 1928)
ÉC *Études Celtiques*
GMW Evans, D. Simon, *A Grammar of Middle Welsh* (Dublin: Dublin Institute for Advanced Studies, 1964)
GPC *Geiriadur Prifysgol Cymru* (University of Wales, 2018) <http://geiriadur. ac.uk/gpc/gpc.html>
IM Honorius Augustodunensis, *Imago mundi*
MWM Huws, Daniel, *Medieval Welsh Manuscripts* (Aberystwyth: National Library of Wales, 2000)
NLW National Library of Wales
RB Red Book of Hergest
RB-A Red Book of Hergest Version A of *Delw y Byd*
RB-B Red Book of Hergest Version B of *Delw y Byd*
RG Luft, Diana, Peter Wynn Thomas, and D. Mark Smith, eds, *Rhyddiaith Gymraeg 1300–1425* (2013) <http://www.rhyddiaithganoloesol.caerdydd. ac.uk>
WB White Book of Rhydderch
ZCP *Zeitschrift für Celtische Philologie*

INTRODUCTION

The Text

This edition presents a selection of chapters from the medieval Welsh encyclopedic text *Delw y Byd* 'Image of the World'. This text is the medieval Welsh translation of Book I of the medieval Latin encyclopedia *Imago mundi* (hereafter referred to as *IM*), composed by Honorius Augustodunensis (*fl. c.* 1090–*c.* 1140).[1] Four distinct versions of the *IM* have been identified, all produced by Honorius, and dated to 1110, 1123, 1133, and 1139.[2] It is the two earliest versions of *IM* that were translated into Welsh, probably sometime in the thirteenth century.[3]

The date of composition places the *IM* in what has sometimes been described as an age of encyclopedism, or the twelfth-century renaissance.[4] The many uncertainties surrounding its origins serve to illustrate its adherence to a wider European intellectual tradition. The author, Honorius Augustodunensis, sometimes referred to as Honorius Inclusus or Honorius of Autun, has proven to be a somewhat elusive figure.[5] The general opinion is that Honorius, whose

---

[1] Although the label 'encyclopedia' is somewhat anachronistic, it has been used of this text, and this type of medieval text, in academic discussion so far, and provides a useful generic label. For a discussion of the anachronistic nature of the term's application to medieval texts, see Arnaud Zucker, 'Introduction', in *Encyclopédire: Formes de l'Ambition Encyclopédique dans l'Antiquité et au Moyen Âge*, ed. by Zucker (Turnhout: Brepols, 2013), pp. 11–28 (p. 12). For a bibliography on medieval encyclopedism, see Benoît Beyer de Ryke, 'Le miroir du monde: un parcours dans l'encyclopédisme médiéval', *Revue belge de philologie et d'histoire*, 81 (2003), 1243–75.

[2] Honorius Augustodunensis, *Imago mundi*, ed. by V. I. J. Flint (Paris, 1982), p. 40.

[3] N. I. Petrovskaia, '*Delw y Byd*: la traduction médiévale en gallois d'une encyclopédie latine et la création d'un traité géographique', *ÉC*, 39 (2013), 257–77 (pp. 276–77). For more, see below, pp. 7–10.

[4] See Bernard Ribémont, *La «Renaissance» du XIIᵉ siècle et l'Encyclopédisme* (Paris: Champion, 2002); note, however, that the same description has also been applied to the thirteenth century; see, in particular, Jacques Le Goff, 'Pourquoi le XIIIe siècle a-t-il été plus particulièrement un siècle d'encyclopédisme?', in *L'enciclopedismo medievale*, ed. by M. Picone (Ravenna: Longon, 1994).

[5] For a biography, see V. I. J. Flint, 'Honorius Augustodunensis of Regensburg', in *Authors of the Middle Ages Historical and Religious Writers of the Latin West*, vol. 2: Nos. 5–6, ed. by C. J. Mews (Aldershot: Variorum, 1995), pp. 89–183. As Richard Southern points out, the mystery was largely Honorius's own doing; *Saint Anselm: A Portrait in a Landscape* (Cambridge: Cambridge University Press, 1990), p. 376; see ibid., n. 9 for references to further discussions; see also Eva Matthews Sanford, 'Honorius, *Presbyter* and *Scholasticus*', *Speculum*, 23 (1948), 397–425. It has even been suggested that he was an Irishman or an Englishman; R. Bauerreiss, 'Zur Herkunft des Honorius Augustodunensis', discussed in Sanford, 'Honorius', p. 401. For the suggestion that Honorius was an Irishman, see also

notable works also include the *Speculum ecclesiae* and the *Elucidarium*, probably spent the latter part of his life in Regensburg, possibly at the *Schottenkloster* or Irish foundation there.[6] His theological outlook appears to reflect the influence of John Scotus Eriugena (ninth century) and Anselm of Canterbury (1033–1109).[7] Indeed, the consensus is that he had spent some time with Anselm in Canterbury, and more generally in England, and it appears that the *IM* at least may have been written during his English period.[8]

The original text of the *IM* is composed of three books, dealing with the physical world, the measurement of time, and history.[9] The first book appears to

---

Southern, *St Anselm and His Biographer* (Cambridge: Cambridge University Press, 1963), pp. 209–17, and Roger E. Reynolds, 'Further Evidences for the Irish Origin of Honorius Augustodunensis', *Vivarium*, 8 (1969), 1–7.

[6] The *Elucidarium*, like the *IM*, exists in a medieval Welsh translation (the *Historia Lucidar*). Both Latin texts have been edited by Jean-Paul Migne in Patrologia Latina 172. For more on the *Elucidarium*, see Valerie Flint's 'The Original Text of the "Elucidarius" of Honorius Augustodunensis from the Twelfth Century English Manuscripts', *Scriptorium*, 18 (1964), 91–94, 'The "Elucidarius" of Honorius Augustodunensis and the Reform in Late Eleventh Century England', *Revue bénédictine*, 85 (1975), 178–89, and 'The Sources of the "Elucidarius"', *Revue bénédictine*, 85 (1975), 190–98. For the Welsh translation, see *The Elucidarium and Other Tracts in Welsh*, ed. by J. Morris Jones and John Rhŷs (Oxford: Clarendon Press, 1894), pp. 3–76, and D. Simon Evans, *Medieval Religious Literature* (Cardiff: University of Wales Press, 1986), pp. 62–63. For more on the *Speculum ecclesiae*, see H. Meyer, 'Zum typologischen Grund der Triumphmetapher im „Speculum Ecclesiae" des Honorius Augustodunensis', in *Verbum et Signum. Beiträge zur mediävistischen Bedeutungsforschung*, ed. by H. Fromm, W. Harms, and U. Ruberg (München: Fink, 1975), pp. 45–58. For more on the Irish monastery at Regensburg, see Pádraig A. Breatnach, 'The Origins of the Irish Monastic Tradition at Ratisbon (Regensburg)', *Celtica*, 13 (1980), 58–77.

[7] Southern, *St Anselm*, p. 377; Loris Sturlese, 'Zwischen Anselm und Johannes Scotus Eriugena: der seltsame Fall des Honorius, des Mönchs von Regensburg', in *Historia Philosophiae Medii Aevi*, ed. by Burkhard Mojsisch and Olaf Pluta, 2 vols (Amsterdam: John Benjamins, 1991), II, pp. 927–51. Whilst there are several authors to whom Honorius's work, and the *IM* in particular, is occasionally misattributed in the manuscript tradition, including Isidore of Seville and Henry of Huntington (e.g. in British Library Royal 13 A xxi), the name of Anselm is prominent. On Honorius and Eriugena, see, for instance, Paolo Lucentini, *Platonismo medievale. Contributi per la storia dell'eriugenismo* (Florence: La nuova Italia, 1980), pp. 40, 45, 56–73; S. Gersh, 'Honorius Augustodunensis and Eriugena'. Remarks on the Method and Content of the "Clavis Physicae"', in *Eriugena Redivius. Zur Wirkungsgeschichte seines Denkens im Mittelalter und im Übergang zur Neuzeit*, ed. by W. Beierwaltes (Heidelberg: Winter, 1987), pp. 162–73; Daniel Yingst, '*Quae Omnia Concorditer Consonant*: Eriugena's Universe in the Thought of Honorius Augustodunensis', in *Eriugena and Creation*, ed. by W. Otten and Michael I. Allen (Turnhout: Brepols, 2014), pp. 427–61. On Honorius and Anselm, see Robert Darwin Crouse, 'Honorius Augustodunensis: Disciple of Anselm?', *Analecta Anselmiana*, 4 (1975), 131–39, and Southern, *Saint Anselm: A Portrait*, pp. 376–81.

[8] *IM*, ed. by Flint, pp. 7–8.

[9] *IM*, ed. by Flint, p. 13; Flint, 'World History in the Early Twelfth Century: The Imago Mundi of Honorius Augustodunensis', in *The Writing of History in the Middle Ages*, ed. by R. H. C. Davis and J. M. Wallace-Hadrill (Oxford: Clarendon Press, 1981), pp. 211–38.

have exercised a particular fascination on medieval audiences and translators, and begat a truly pan-European family of texts. Over one hundred manuscripts of the Latin text survive, excluding fragments, as well as medieval translations into Spanish, French, Italian, Middle English, and Welsh.[10] *Delw y Byd* thus appears to be representative of the tendency, noted by Brynley F. Roberts in relation to medieval Welsh translated works, of making the most popular foreign material available and as accessible as possible, to a Welsh audience.[11] The objective of this edition is to provide students of medieval Welsh with a glimpse into the type of encyclopedic material available to medieval Welsh audiences.

Previous Editions

The existing edition of *Delw y Byd* is *Delw y Byd (Imago Mundi)*, ed. by Henry Lewis and P. Diverres (Cardiff: University of Wales Press, 1928), henceforth referred to as *DB*. This edition presents the texts of three of the five surviving medieval manuscripts; the other two were unknown at that time. The editors also reproduce a Latin text in parallel. Their Latin text is taken from Honorius Augustodunensis, *De imagine mundi libri tres*, ed. by Jean-Paul Migne,

[10] Flint, 'Honorius Augustodunensis', pp. 165–67. Note that Flint's list is incomplete and does not include the numerous fragments and extracts that survive elsewhere; missing, for instance, is a reference to Exeter, Cathedral Library MS 3514, a Welsh manuscript which contains Book I of the *IM*; for more on this manuscript, see below, 4, 8, 11 n. 52, and 14-15. For an overview of other vernacular translations of the *IM*, see R. E. Kaske et al., *Medieval Christian Literary Imagery: A Guide to Interpretation* (Toronto, Buffalo and London: University of Toronto Press, 1988), pp. 189–91. Even where there appears to be a lack of vernacular translation, familiarity with the *IM* is attested. An example of this is the Norse tradition; see, for instance, Svanhildur Óskarsdóttir, 'Prose of Christian Instruction', in *A Companion to Old Norse-Icelandic Literature and Culture*, ed. by Rory McTurk (Malden MA, Oxford and Victoria: Blackwell, 2005), pp. 338–53 (p. 342); for a broader discussion, see Rudolf Simek, *Heaven and Earth in the Middle Ages: The Physical World Before Columbus*, trans. by Angela Hall (Woodbridge: Boydell, 1996). These are translations in the very broad sense of the word, and their fidelity to the original ranges from very faithful to a complete restructuring and incorporation into a new encyclopaedic format (e.g. Gossouin de Metz's French *Image du monde* and its Italian and Middle English translations). For a comparative discussion of the Welsh and French versions, see N. I. Petrovskaia, 'Mythologizing the Conceptual Landscape: Religion and History in *Imago Mundi*, *Image du Monde* and *Delw y Byd*', in *Landscape and Myth in North-Western Europe*, ed. by M. Egeler (Turnhout: Brepols, 2018). For a list of known manuscripts of the Latin and all vernacular versions, see N. I. Petrovskaia and Kiki Calis, 'Images of the World: Manuscripts Database of the Imago Mundi Tradition' <https://imagomundi.hum.uu.nl> (2019) [accessed 17 July 2019].

[11] Brynley F. Roberts, '*Ystoriaeu Brenhinedd Ynys Brydeyn*: A Fourteenth-Century Welsh Brut', in *Narrative in Celtic Tradition*, ed. by Joseph F. Eska (Hamilton, NY: Colgate University Press, 2011), pp. 217–27 (p. 217); Alexander Falileyev has justly drawn attention to the importance of *Delw y Byd* calling for a new edition; '*Delw y Byd* Revisited', *Studia Celtica*, 44 (2010), 71–78 (esp. p. 73).

Patrologiae Cursus Completus, Series Latina 172 (Paris, 1895).[12] They also follow Migne's chapter numbering. A new edition of the Latin text has since appeared: *Imago mundi*, ed. by V. I. J. Flint, Archives d'histoire doctrinale et littéraire du moyen Âge 49 (Paris, 1982). This edition differs in its chapter divisions from the earlier ones. I have chosen to follow Flint's chapter divisions in the present text, but, as an acknowledgment of the importance of *DB*, I reproduce their chapter numbers consistently in square brackets in my discussion, text, and glossary.[13] I hope that this approach will make reference to earlier editions and discussions easier. I have also provided section headings reflecting the structure of the Latin text. A full chapter list (giving both sets of chapter numbers) is provided on pp. 43–45 below. The present edition reproduces sixty-seven of the 136 chapters of the Welsh text, as well as the two introductory letters. For seventeen of the chapters, variants from all manuscripts are provided. The chapter list below provides an overview of which chapters occur in which manuscripts of the Welsh text, and marks those chapters reproduced in the present edition.

Organization of the Welsh Text and its Latin Source

The structure of Book I of the *IM* corresponds to the four elements, which Honorius conceives as concentric spheres encompassing all creation. Earth (geography), Water (oceans, rivers, weather phenomena), Air (winds), and Fire (heavenly spheres, astronomy, the zodiac) are all treated in order[14] (see Figure 3 accompanying the text on p. 48 below). The geographical section of the text ('Earth') corresponds to the view of the world also transmitted through the medieval world maps, the *mappae mundi*, such as the Hereford Map and the Psalter Map, as well as the Exeter Map and the Corpus Map, the last of which accompanies a text of *IM* preserved in that manuscript.[15] This type of geographical ordering of the world is known as the T-O map. Ultimately based on classical sources, and popularized by Isidore of Seville (c. 560–636), this

---

[12] Based on Roger Wilmans's edition, 'Ex Honorii Augustodunensis summa totius et imagine mundi', Monumenta Germaniae Historica Scriptores XII (Hannover, 1852), pp. 125–34.

[13] For a more detailed discussion, see pp. 29–30 below.

[14] Whilst it was common for medieval thinkers to identify four elements, as Honorius does here, some added ether as a fifth element; see A. Pablo Iannone, *Dictionary of World Philosophy* (London and New York: Routledge, 2001), s.v. element. Use of coloured initials and capitals in the manuscripts of the Welsh translation of the text shows that the compilers were aware of the text's structure; for more on use of capitals and coloured initials in the manuscripts, see pp. 31–32 below.

[15] Images of the former two are available online at the Hereford Cathedral Mappa Mundi website <http://www.themappamundi.co.uk> [accessed 30 March 2016] and <http://www.bl.uk/collection-items/psalter-world-map> [accessed 30 March 2016]; for images of the latter maps, see N. I. Petrovskaia, *Medieval Welsh Perceptions of the Orient* (Turnhout: Brepols, 2015), plates 1 and 2.

scheme presents the world as a circle, divided by a T-shape combination of the Mediterranean, the Don, and the Nile into three parts or continents: Asia, Europe, and Africa.[16] Such maps were conventionally oriented towards the East (see Figure 6 on p. 50 below). It can be argued that the *IM* also follows this organizational principle, as the narrative starts with a description of Paradise, and the easternmost of the lands of Asia, before progressing to a discussion of the rest of Asia, followed by Europe and Africa.

The type of geographical information contained in this text, and echoed visually in the *mappae mundi*, was not intended as an aid to navigation or travel, but rather, like the information regarding the beasts and the spheres, constituted a tradition handed down from the authorities of antiquity and was intended as an aid to the understanding of the world as a system.

Surrounding the Earth was the sphere of Water (chapters 38[38]–57[52]). This section deals with the concepts of the oceans, fresh and salt water, fish inhabiting the sea, and tempests. This is followed by the section on Air (chapters 58[53]–71[66]), which concerns the twelve winds, aerial phenomena such as the rainbow, rain, fog, and the clouds. Finally, the section devoted to the element of Fire (chapters 72[67]–147[140]) deals with astronomy, according to the Ptolemaic (geocentric) system, culminating in a description of the heaven of heavens. The system described in the text is thus finite, self-contained, and harmonious.[17] To illustrate the type of information provided in *Delw y Byd*, the present edition reproduces selected chapters from each section, including the chapters at the end and beginning of sections, which convey the flow and structure of the text.

As Flint notes, the sources of information in *IM* which can be identified are all standard authorities: Pliny, Solinus, Orosius, Macrobius, Isidore, Martianus Capella, Bede, Rabanus Maurus, Helpericus, Pseudo-Bede, and Pseudo-Alcuin.[18] What can be added to this observation is that the main source for the *IM* appears to have been Isidore. Even a cursory glance at the notes to Flint's edition shows that there are almost no instances where sources identified by her for individual chapters do not include at least one of Isidore's works. Indeed, in some of the very few cases where Flint does not give Isidore as the source, the relevant information can nevertheless be identified in his works; for instance,

---

[16] David Woodward, 'Medieval *Mappaemundi*', in *History of Cartography I: Cartography in Prehistoric, Ancient, and Medieval Europe and the Mediterranean*, ed. by J. B. Harley and David Woodward (Chicago and London: University of Chicago Press, 1987), pp. 286–370 (pp. 301–02).

[17] C. S. Lewis gives a comparison between the medieval and the modern astronomical conception of space in *The Discarded Image: An Introduction to Medieval and Renaissance Literature* (Cambridge: Cambridge University Press, 1994), p. 99.

[18] *IM*, ed. by Flint, p. 13.

for Chapter 87[82] 'On man', a source not noted by Flint is Isidore's *Etymologies*, III. 23.[19]

The *IM* appears to have been deemed pertinent to a range of disciplines, as attested by the variety of the manuscript contexts within which it can be found. Extracts survive in manuscripts of historical, medical, and political, as well as scientific interest.[20] The manuscript context of the Welsh translations is discussed in greater detail below.

Manuscripts of *Delw y Byd*

Before proceeding to a discussion of the manuscript tradition of *Delw y Byd* it is necessary to take a brief look at the manuscript tradition of the *IM*, to which it is tightly linked. As mentioned above, four distinct versions of the *IM* have been identified, all produced by Honorius, and dated to 1110, 1123, 1133, and 1139.[21] It appears that Honorius progressively updated the text by adding extra information to existing chapters and inserting additional chapters in all three books. The earliest version is therefore also the shortest. This early version appears to have particular links with Britain, and there is a suggestion that Honorius may have composed his text during his stay in England, possibly at Canterbury (hence the occasional attribution of the text to St Anselm). The earliest manuscripts of the 1110 version are English. They are:

Cambridge, Corpus Christi College MS 66 (end of the twelfth century, Sawley, Lancashire)

London, British Library MS Royal 13 A xxi (fourteenth century, Hagneby, Lincolnshire)

Oxford, Bodleian Library, MS Rawlinson B 484, ff. 1r–6v (end of the twelfth century, Britain)[22]

These manuscripts are important to us because their version of the text is the

[19] Throughout the introduction, I provide titles for the chapters discussed based on the Latin titles provided in Valerie Flint's edition. C. S. Lewis observes that it is in Isidore's work that the process of the creation of what Lewis calls 'pseudo-Zoology' (a peculiarly medieval attitude to the study of the animal world) can be observed; *The Discarded Image*, p. 148. For a discussion of Isidore's geography, see Andy Merrills, 'Geography and Memory in Isidore's *Etymologie*', in *Mapping Medieval Geographies: Geographical Encounters in the Latin West and Beyond, 300–1600*, ed. by Keith Lilley (Cambridge: Cambridge University Press, 2013), pp. 45–64.

[20] For a list of *IM* manuscripts and information on their content see the 'Images of the World' database.

[21] *IM*, ed. by Flint, p. 40.

[22] The date relates to these folios only; for more on this composite manuscript, see N. I. Petrovskaia, 'The Travels of a Quire from the Twelfth Century to the Twenty-First: The Case of Rawlinson B 484, fols. 1–6', in *Middle English Texts in Transition: A Festschrift in Honour of Toshiyuki Takamiya*, ed. by M. Driver, L. Mooney and S. Horobin (York: York Medieval Studies, 2014), pp. 250–67.

basis for one of the two Welsh translations. The other translation was based on the 1123 version. There is a third possible 'translation', the basis for which is difficult to identify since it survives only in one chapter fragment, but it may also belong to the translation based on the 1110 version (this is discussed in greater detail below).[23] The exact relationship between these versions will be discussed further, and the relevant stemmata presented, below.

Turning to the Welsh text, there are five medieval manuscripts containing six texts of *Delw y Byd*.[24] These are presented in Table 1. One manuscript, the Red Book of Hergest (henceforth RB), contains copies of both versions of the text. None of the manuscripts contain a text corresponding to the whole of *IM* Book I, but it appears that a shorter version, ending at Chapter 81[76] 'Of Saturn', was in circulation, since three of our manuscripts end the consecutive text at that chapter (in two cases this is followed by miscellaneous additions, discussed below). There is evidence that a complete translation of Book I of the *IM* was also in circulation at some point, however, as the acephalous Peniarth 17 text continues to the end of Book I of the *IM* as we know it, Chapter 147[140] 'Heaven of Heavens'. One of the surviving manuscripts, Rawlinson B 467, contains extracts of the text, rather than fragments.[25]

The six surviving medieval Welsh texts represent two, and possibly three, different versions of the text, stemming from two separate translations of differing versions of the *IM*, the 1110 and the 1123.[26] We will call the first 'Version A' and the second 'Version B'.[27] These are primarily distinguished by the passages added to Version B, such as, for instance, information about the magnet added to Chapter 12[13] 'On Beasts'.[28] Chapter 26[28] 'On Italy', represents an additional indicator of the difference, as the different Welsh versions present the elements of that passage in a different order.[29] The Peniarth

---

[23] See pp. 20–22 below.

[24] The Welsh manuscript tradition for *Delw y Byd* is discussed in Petrovskaia, '*Delw y Byd*', and Petrovskaia, *Medieval Welsh Perceptions of the Orient*, pp. 7–15. The two versions of the text present in the RB are discussed in greater detail below, pp. 11, 15, 18–23.

[25] For more on this manuscript and the extracts it contains, see pp. 12–13 and 17–18 below.

[26] In Petrovskaia, '*Delw y Byd*', p. 272, I assigned both of the Welsh translations to the 1110 version of *Delw y Byd*. Alexander Falileyev has pointed out that the manuscripts of Version B contain additions belonging to the 1123 version of the *IM*; Falileyev, '*Delw y Byd* Revisited', pp. 75–76. The two separate versions were identified in *DB*. Note that (*pace* Falileyev), the editors do not identify the section of the RB text printed as their text 'C' as a separate version. Its designation is used to identify it, since it is separated from the rest of the corresponding RB text, which is used to fill in the gap of the acephalous Peniarth 17 as 'Text A'; *DB*, p. xv; Falileyev, '*Delw y Byd* Revisited', pp. 73, 74.

[27] The designation corresponds to the use of A and B for different version of *Delw y Byd* established in *DB*.

[28] For details of the additions, see *IM*, ed. by Flint, p. 37–38. For the magnet, see the note to Chapter 12[13] on p. 86 below.

[29] The Peniarth 17 and Red Book (cols 975–99) transpose the order of the passages in this

17 text and one of the RB texts (cols 975–99) represent Version A. The other RB text (cols 502–16), and the White Book, Rawlinson B 467, and Philadelphia texts, represent Version B (see list in Table 1). The possible third version, which may represent a variant on Version A, is preserved in the form of an addition to the Version B text in the RB, at col. 516, formed of a group of chapters presented in an unusual order: 74[69] 'The Moon', 77[72] 'The Sun', and 67[62]–69[64] (on the rain, clouds, and smoke). Preliminary examination suggests that this text is closer to Version A of *Delw y Byd* and henceforth I use the label A-516 for this fragment. A-516 is included separately in the manuscript stemma presented as part of the discussion of manuscript relationships in the section 'Relationships Between the Texts' below, and the details are discussed further in the section on structure.[30] There also survives a Welsh manuscript of the Latin text: Exeter, Cathedral Library MS 3514 (Whitland, second half of the thirteenth century).[31] It contains a copy of the 1123 version of the *IM*, the same version of the text as that from which the Welsh Version B was translated.[32]

The following table contains details of the manuscripts of *Delw y Byd*, listed in chronological order, and provides information on the sections of the text they contain.[33]

Table 1: The Welsh manuscripts and their contents

MS details	Folios	DyB version and chapters[34]
Aberystwyth, NLW, MS Peniarth 17 (thirteenth century)	pp. 17–26[35]	A Version Contains: 53[48]–147[140] Omitting: 78[73], 120[114]–125[119], 138[132]

chapter, while the White Book and Red Book (cols 502–16) texts abridge, but maintain the order as it stands in the majority of Latin manuscripts. The inverse order found in Peniarth 17 and the Red Book (cols 975–99) is found to my knowledge in only one manuscript of the Latin text, Rawlinson B 484. For a detailed discussion, see Petrovskaia, 'Delw y Byd', p. 267–70 and 273–75, and Petrovskaia, 'Travels of a Quire'.

[30] See below, pp. 15 and 22–23.

[31] Julia Crick identifies one of the scribes in this manuscript as active *c.* 1266 and another *c.* 1285, with the *IM* scribe active somewhat earlier than the 1266 scribe; 'The Power and the Glory: Conquest and Cosmology in Edwardian Wales (Exeter Cathedral Library, MS. 3514)', in *Textual Studies: Cultural Texts*, ed. by Elaine Treharne and Orietta da Rold (Cambridge: D. S. Brewer, 2010), pp. 21–42 (pp. 24, 33, 40).

[32] It contains the 1123 additions noted by Flint but not the later additions; *IM*, ed. by Flint, p. 37.

[33] For more on the manuscripts, see *MWM*, pp. 58–60.

[34] For a detailed discussion of the chapters included and omitted, see below, pp. 43–45.

Aberystwyth NLW, MS Peniarth 5, White Book of Rhydderch (c. 1350)	ff. 2r–4r	B Version Contains: 12[13]–81[76] Omitting: 44[43]–47[44], 50–51[46], 53[48]–57[52], 71[66]–73[68], 78[73]
Oxford, Jesus College MS 111, Red Book of Hergest (1382× c. 1400)	1 21v–125r (cols 502–16)[36]	B Version Contains: title, 1[1]–81[76] Omitting: 4[4], 44[43]–47[44], 50–51[46], 53[48]–57[52], 71[66]–73[68], 78[73] A Version (fragments) Additional chapters: 74[69], 77[72], 67[62]–69[64]
Oxford, Jesus College MS 111, Red Book of Hergest (1382× c. 1400)	242v–248v (cols 975–99)	A Version Contains: Introductory letters, 1[1]–81[76], an additional original chapter, and 88[83] Omitting: 4[4], 53[48]–56[51], 71[66], 78[73]
Oxford, Bodleian Library, MS Rawlinson B 467 (c. 1400)	ff. 70v–72v[37]	B Version Contains: 38[38]–41[41] and 59[54]–63[58]
Library Company of Philadelphia, MS 8680.O (1382× c. 1400)[38]	ff. 1r–2r	B Version Contains: 11[12]–12[13] and 18[19]–24[26]

Two further surviving manuscripts, both early modern in date, contain (parts of) *Delw y Byd*. Further investigation is needed to establish their place in the tradition and their relation to the earlier versions, and they are not included in the present edition. The earlier of the two is National Library of Wales MS 5267B, which contains a copy of *Delw y Byd* on ff. 1–10.[39] Formerly considered

---

[35] Note that this manuscript is paginated, not foliated.

[36] Note that the column range given in *DB*, p. 85 (as 'col. 502–06'), is a typographical error.

[37] *MWM*, pp. 41, 43, 60.

[38] One of the scribes active in this manuscript is Hywel Fychan; *MWM*, p. 60. For more on this manuscript, see also Gruffydd Aled Williams, 'Mwy am Lawysgrif Gymraeg yn U.D.A.: The Public Library Company of Philadelphia, Llsgr. 8680.O', *Llên Cymru*, 34 (2011), 248–50, and Ben Guy, 'A Welsh Manuscript in America: Library Company of Philadelphia, 8680.O', *National Library of Wales Journal*, 36 (2014), 98–123.

[39] For a description and discussion of the manuscript, see Rebecca Try, 'NLW MS 5267B;

to have been produced in the seventeenth century, its date has recently been revised to the first half of the fifteenth century (possibly 1438).[40] The other early modern witness to the text is Cardiff, Central Library, 2.83 (c. 1550), pp. 25–26, containing chapters 5[5] to 8[8] of the text.[41] The *Delw y Byd* fragment preserved in this manuscript is reproduced at the end of *DB* and is also briefly discussed by Alexander Falileyev, who has identified it as an 'Early Modern Welsh adaptation' of Version B.[42]

The Welsh Translation: Manuscript Context and Audiences

Two audience types can be distinguished for *Delw y Byd*. The first is those for whom the text was originally translated, and it is not necessarily the case that the intended audience was the same both times the text was translated. The second type is the audience(s) of the manuscripts in which our texts survive. The discussion that follows focuses on the latter audience type. The contexts in which our text is found suggest that the different copies may have been intended for different audiences, and the text perceived at various times as belonging, or perhaps relevant, to different genres.

More has been written on the audiences of the RB, created for Hopcyn ap Tomas (c.1330–after 1403), and the White Book of Rhydderch (henceforth WB), created for Rhydderch ab Ieuan Llwyd (c. 1325–1400), than of the other manuscripts in which *Delw y Byd* survives.[43] Both patrons belonged to the *uchelwyr* or minor nobility, and both books are compilations of miscellaneous Welsh-language texts. The stress, however, is different in the two compilations. As Daniel Huws notes, in the design of the RB '[o]bviously deliberate exclusions were law and religion'.[44] The immediate context of Version B of *Delw y Byd* in the RB is historical: it follows the Welsh translation of the Latin *Pseudo-Turpin Chronicle*, a pseudo-historical account of Charlemagne's exploits in Spain, and is followed by a short chronicle text probably of later fourteenth century

---

A Partial Transcription and Commentary' (unpublished MPhil dissertation, Cardiff University, 2015).

[40] See *MWM*, p. 61, and Try, 'NLW MS 5267B', p. vi, with reference to Daniel Huws's unpublished description.

[41] For a description of the manuscript, see J. Gwenogvryn Evans, *Report on Manuscripts in the Welsh Language*, 2 vols (London: for H.M.S.O. by Eyre and Spottiswoode, 1898–1905), II. 1, pp. 104–06.

[42] *DB*, pp. 112–13; Falileyev, '*Delw y Byd* Revisited', p. 74.

[43] See, for instance, *MWM*, pp. 227–68; Catherine McKenna, 'Reading with Rhydderch: Mabinogion Texts in Manuscript Context', in *Language and Power in the Celtic World: Papers from the Seventh Australian Conference of Celtic Studies*, ed. by Anders Ahlqvist and Pamela O'Neill (Sydney: University of Sydney, 2011), pp. 205–30.

[44] *MWM*, p. 82.

composition, which appears to end with the year 1353,[45] and the *Hwsmonaeth* (a translation of Walter of Henley's francophone treatise on husbandry).[46] The aspect of the script is different between the *Delw y Byd* and the subsequent text, suggesting that there was a pause of some time between the completion of one text and the addition of the other. This could be interpreted as suggesting that the scribe had searched for a continuation of his *Delw y Byd* text (faced, perhaps, with an incomplete exemplar), or that in the interim he had forgotten that the text was incomplete (where the exemplar would have continued).[47] Version A is copied between *Diarhebion* (a collection of proverbs) and *Brenhinedd y Saeson* (a historical text).[48] Both versions of the text in the RB are copied by Hywel Fychan, as are the texts surrounding Version A and the texts following Version B.[49] The *Pseudo-Turpin Chronicle* is in the hand of a different scribe, designated Scribe A by Daniel Huws, but it is worth noting that Huws acknowledges the difficulty in distinguishing between Scribe A and Hywel Fychan, particularly regarding the text of *Delw y Byd* Version B (ff. 121v–125).[50] The fact that Hywel copied both *Delw y Byd* texts in the RB suggests that these might have been perceived as distinct entities.[51]

The historical and scholarly aspect of the RB setting is echoed in Peniarth 17, where our text is accompanied by a biography of Gruffudd ap Cynan, and the *Diarhebion* or proverb collection.[52] This contrasts with the context (exclusively

---

[45] RB ff. 125r–v. Given the decoration at the end of the line, this appears to be the end of the chronicle. However, it should be noted that the rest of the page is obscured by the supporting vellum leaf pasted on to it. It appears to have been blank; see Evans, *Report*, II, p. 2 n.

[46] *Ystorya de Carolo Magno o Llyfr Coch Hergest*, ed. by Stephen J. Williams (Cardiff: University of Wales Press, 1968); *Welsh Walter of Henley*, ed. by A. Falileyev (Dublin: Dublin Institute for Advanced Studies, 2006).

[47] The RB-B text does not have a concluding phrase, suggesting the possibility that the text was perceived as incomplete; see Table 2 and accompanying discussion, p. 19 below.

[48] *Diarhebion*, ed. by Richard Glyn Roberts (Aberystwyth: CMCS, 2013), pp. 107–29; *Brenhinoedd y Saeson = 'The Kings of the English'*, A.D. 682–954, ed. and trans. by David N. Dumville (Aberdeen: University of Aberdeen, 2005).

[49] Daniel Huws, 'Llyfr Coch Hergest', in *Cyfoeth y Testun. Ysgrifau ar Lenyddiaeth Gymraeg yr Oesedd Canol*, ed. by I. Daniel, M. Haycock, D. Johnston, and J. Rowland (Cardiff: University of Wales Press, 2003), pp. 1–30 (p. 5). See also the TEI header for the manuscript on Diana Luft, Peter Wynn Thomas, and D. Mark Smith, eds, *Rhyddiaith Gymraeg 1300–1425* (2013) <www.rhyddiaithganoloesol.caerdydd.ac.uk> [accessed 26 October 2015], henceforth referred to as *RG*; for more on Hywel Fychan, see *MWM*, pp. 79–83, and Gifford Charles-Edwards, 'The Scribes of the Red Book of Hergest', *National Library of Wales Journal*, 21 (1980), 246–56 (pp. 250–53).

[50] Huws, 'Llyfr Coch Hergest', p. 12.

[51] By way of comparison, cf. the discussion of two co-existing versions of the introduction to *Bown* in *Selections from Ystorya Bown o Hamtwn*, ed. by Erich Poppe and R. Reck (Cardiff: University of Wales Press, 2009), pp. 14–15.

[52] For an edition of the former, see *Historia Gruffud vab Kenan*, ed. by D. Simon Evans (Cardiff: University of Wales Press, 1977); for an edition of the latter, see Henry Lewis, 'Y Diarhebion ym Mheniarth 17', *BBCS*, 4 (1927–1929), 1–17. Compare the apparent political

religious) in which the text is found in the WB.[53]

While the RB compilers appear to have avoided religious material, the text of *Delw y Byd* preserved in the WB, a mid-fourteenth-century Strata Florida production, was copied by the scribe (designated scribe A by Daniel Huws) responsible for the first four quires of the manuscript, and thus for the texts of *Proffwydoliaeth Sibli Ddoeth*, *Gwyrthiau Mair*, and *Efengyl Ieuan* (all texts of religious nature).[54] It could be argued that our text, with its focus on describing creation, was perceived by the compiler of this section of the WB also to be religious in nature.[55] On the basis of differences in layout and contents, Huws has suggested that the first four quires of the WB (*Delw y Byd* is in the first of these) were originally intended as a different manuscript, 'written presumably for use by a cleric'.[56] Huws suggests that 'these four quires were added to the book as an afterthought. Rhydderch, or someone, decided to mix secular and religious reading in one book (Peniarth 7 and 14 offer precedents). Perhaps the religious texts were placed at the beginning of the book in a spirit of piety'.[57] In this context, the *Delw y Byd* text appears to have been regarded as a religious text. As suggested above, its subject matter describing the Creation, not to mention the focus of its geographical section on biblical geography, may well account for such a reading of the text.

The composite nature of the Philadelphia manuscript permits very little to be gleaned of the original (intended) readership of our text.[58] The *Delw y Byd* quire was added to the Hywel Fychan material of the manuscript in the second half of the sixteenth century, and thus cannot be associated with the activities of that scribe. However, given that the *Delw y Byd* quire appears to have originally belonged with the quire containing a calendar (albeit one not composed for liturgical use),[59] it can be tentatively suggested that, as with the WB, its original manuscript context points to a religious environment and a monastic audience.

Finally, the scholarly or scientific reading of *Delw y Byd* has its apogee in its inclusion in a medical collection in Rawlinson B 467.[60] We can thus observe

---

application to which the *IM* is put in the Exeter manuscript, where it is found alongside historical texts, contextualizing Welsh history on a global scale. For discussion, see Crick, 'The Power and the Glory'; for more on the uses of Honorius's text in political and historical contexts, see Flint, 'World History'. See also Richard Glyn Roberts, 'Dalen Olaf Llawysgrif Peniarth 17', *Dwned*, 9 (2003), 37–42.

[53] See discussion in *MWM*, cited below.

[54] *MWM*, pp. 231, 234.

[55] For more information on the relationship between geography and spiritual education, see David Woodward, 'Reality, Symbolism, Time, and Space in Medieval World Maps', *Annals of the Association of American Geographers*, 75 (1985), 510–21.

[56] *MWM*, pp. 244–45, 253–54.

[57] *MWM*, p. 254.

[58] Guy, 'A Welsh Manuscript in America', pp. 10–11.

[59] Guy, 'A Welsh Manuscript in America', p. 10.

[60] Morfydd E. Owen, 'The Medical Books of Medieval Wales and the Physicians of

that the material could be applied to a number of distinct fields of knowledge. This suggests that the Welsh audiences, compilers, and scribes were engaged in seeking out and collecting a wide variety of texts relevant to the chosen subject, whether religion, history, or medicine.

This brief overview of the manuscript context within which the *Delw y Byd* text is found demonstrates not only considerable adaptability on the part of the text itself, but also a great flexibility on the part of the Welsh redactors and audiences. The compilers of our manuscripts were less concerned with rigid distinctions between disciplines and more with enriching their collections of texts in the best possible manner. In this process, no violence appears to have been done to the text itself, if one is to discount the act of extracting passages for the medical collection in Rawlinson B 467.

One might argue that this flexibility is a function of the original material and that the Welsh translators merely transplanted the information carried by the Latin text into a different linguistic milieu.[61] The Welsh translators' approach to this text corresponds to the attitude expressed in a much-quoted passage by Gruffud Bola (*fl.* 1265–1282) in his translation of the Athanasian Creed:

> Vn peth hagen a dylyy ti y wybot ar y dechreu, pan trosser ieith yn y llall, megys Lladin yg Kymraec, na ellir yn wastat symut y geir yn y gilyd, a chyt a hynny kynnal priodolder yr ieith a synnvyr yr ymadravd yn tec. Vrth hynny y troes i weitheu y geir yn y gilyd, a gveith ereill y dodeis synnvyr yn lle y synnvyr heruyd mod a phriodolder yn ieith ni.[62]

This reflects the dominant medieval attitude towards translation, based on the tradition founded by St Jerome, author of the Vulgate Latin translation of the Bible, whose statement that he aimed to *Non verbum e verbo, sed sensum exprimere de sensu*, 'to render not word for word but sense for sense', is quoted almost verbatim.[63]

Myddfai', *The Carmarthenshire Antiquary*, 31 (1995), 34–44 (p. 35); Diana Luft, 'Ansoddau'r Trwnc: A Welsh Uroscopic Tract', *ZCP*, 58 (2011), 55–86 (p. 60). See also pp. 25–26 below.

[61] This suggestion is supported by the wide range of manuscript contexts, ranging from scientific to historical to devotional, in which the Latin text is found; see reference in n. 20 above.

[62] Henry Lewis, 'Credo Athanasius Sant', *BBCS*, 5 (1930), 193–203 (p. 196); 'When one translates one language into another, for example Latin into Welsh, you should know that it is not always possible to substitute one word for another and retain both the idiom and sense of the expression. Therefore I have sometimes translated word for word, but at other times I have rendered sense for sense according to the nature and idiom of our language', trans. by Brynley Roberts, 'Ystoriaeu Brenhinedd', p. 218. See also Ceridwen Lloyd-Morgan, 'French Texts, Welsh Translators', in *The Medieval Translator 2*, ed. by Roger Ellis (London: Centre for Medieval Studies, Queen Mary and Westfield College, University of London, 1991), pp. 45–63 (p. 56).

[63] Jerome, 'Epistola LVII. Ad Pammachium; De optimo genere interpretandi', in *Sancti Hieronymi Stridonensis Presbyteri Opera Omnia I*, ed. by J.-P. Migne (1845), cols 568–79 (col. 571), my translation. For a discussion of medieval translation practice and further

The awareness of the inevitability of the loss of one aspect or another of the original in translation is displayed particularly in the translators' approach to tackling the multiple etymological formulae present in the text.[64] The translators' goal appears to have been to transmit the information carried by the text into a Welsh-language environment, and to incorporate it into the corpus of material available to Welsh intellectuals. As such, handy and easily manageable versions of the text became available for use for a variety of purposes, including religious, medical, historiographical, and political.

Relationships Between the Texts

In previous publications I have presented a stemma which omitted the Philadelphia and Rawlinson manuscripts.[65] The stemmata below incorporate these and introduce a correction to the derivation of Version B.[66] In addition to manuscripts of the Welsh translation, they include the related Latin manuscripts. The Version A stemma includes the three Latin manuscripts introduced above, and the Version B stemma includes the only known Latin manuscript of Welsh provenance.[67] Whilst Version A of *Delw y Byd* derives from the text of which a fragment is preserved in Rawlinson B 484, it has not so far been possible to identify the Latin source manuscript for the second Welsh translation (Version B).[68] The Exeter manuscript, which contains the 1123 Latin version of the text, is included because the manuscript is of Welsh provenance. The distinction between texts in Welsh and texts in Latin is made in the stemmata by underlining the Welsh-language manuscripts.

references, see N. I. Petrovskaia, '*Translatio* and Translation: The Duality of the Concept from the Middle Ages to the Early Modern Period', 同志社大学英語英文学研究 = *Doshisha Studies in English*, 99 (2018), 115–36 (pp. 120–24).

[64] For a detailed discussion, see N. I. Petrovskaia, 'La disparition du *quasi* dans les formules étymologiques des traductions galloises de l'*Imago Mundi*', in *La Formule au Moyen-Âge*, ed. by E. Louviot (Turnhout: Brepols, 2012), pp. 123–41.

[65] Petrovskaia, '*Delw y Byd*', pp. 272, and 'La disparition du *quasi*', p. 128.

[66] The fragments in these manuscripts have also been identified as belonging to Version B by Brynley F. Roberts, 'Un o lawysgrifau Hopcyn ap Thomas o Ynys Dawe', *BBCS*, 22 (1967), 223–28 (p. 225); see also Falileyev, '*Delw y Byd* Revisited', p. 73.

[67] See above, pp. 6–8; for a detailed discussion, see Petrovskaia, '*Delw y Byd*', pp. 263–65.

[68] The manuscripts of the 1123 version listed by Flint are: Brussels BR 10862–5; BL Add. 38665 and Cotton Cleopatra B.IV; Paris, BnF, lat. 11130 and 15009; Bibl. Arsenal 93(B); Bibl. Mazarine, MS 708; *IM*, ed. by Flint, pp. 36–37. To these can also be added the fragment in Brussels, BR 2419–31; Wouter Bracke, 'Pomponius Mela, Étienne de Byzance, Honorius d'Autun et le MS. de Bruxelles, BR 2419–31', *Revue belge de philologie et d'histoire*, 81 (2003), 1075–81.

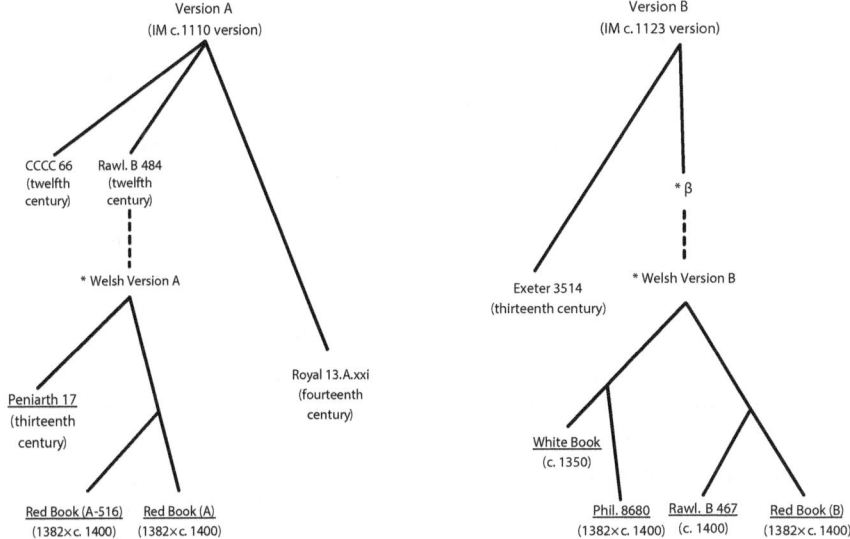

Figure 1: Relationship between *Imago mundi* and *Delw y Byd* manuscripts

The stemmata above are provisional, as the definition of a more precise set of relationships between the four Welsh texts belonging to Version B, beyond the simple statement that these four manuscripts appear to contain copies of the same translation, is hampered by the small size of the fragments in the Rawlinson and Philadelphia manuscripts. Nevertheless, some brief observations can be made here. It is worth noting that whilst they are very close, the Version B texts of *Delw y Byd* preserved in the RB and the WB are not identical. The texts of Version B preserved in the RB and the WB appear to share a common ancestor, a situation similar to that observed for other texts which have copies preserved in both of these manuscripts.[69] In addition to the variation which will become apparent in the discussion below of the affiliation of the Philadelphia and Rawlinson fragments, the most striking example is perhaps in the treatment of place-names, which are almost consistently spelt differently. To give but one example, the list of regions of Spain in Chapter 28[30] 'Spain' is given in RB-B as *Daraconia*, *Karthago*, *Lucitania*, *Gallicia*, *Betica*, *Tignitania*, but in the WB

---

[69] See, for instance, Jenny Rowland, 'The Manuscript Tradition of the Red Book *Englynion*', *Studia Celtica*, 18 (1983), 79–95; Peter Wynn Thomas, 'Cydberthynas y pedair fersiwn ganoloesol', in *Canhwyll Marchogyon: Cyd-destunoli 'Peredur'*, ed. by Sioned Davies and P. W. Thomas (Cardiff: University of Wales Press, 2000), pp. 10–49; Erich Poppe and R. Reck, 'Rewriting *Bevis* in Wales and Ireland', in *Sir Bevis of Hampton in Literary Tradition*, ed. by Jennifer Fellows and Ivana Djordjević (Cambridge: Boydell & Brewer, 2008), pp. 37–50 (p. 38); *Selections*, ed. by Poppe and Reck, p. 14; also *MWM*, pp. 246, 255.

these are given as _Terragonia_, a _Chartago_, a _Llucitammia_, a _Gallicia_, _Betica_, _Tyngnyttrania_ (the differences are underlined).[70] The variation cannot be explained merely through spelling differences, and it is possible that it might be due to a scribe at some point in the transmission of the text correcting the reading according to a Latin text he had at hand. More work needs to be done on establishing the exact relationship between these two Welsh Version B texts and the reasons for these divergences.

Whilst the Philadelphia fragment appears to be closer to the WB, the Rawlinson extracts show more affinity with the RB-B text. Since there are few discussions of these manuscripts, this is treated in greater detail here.

Although, according to Alexander Falileyev, the Philadelphia text is sometimes closer to the RB and sometimes to the WB text, a detailed examination shows that the Philadelphia manuscript is in agreement with the WB against the RB-B version in a number of significant variants.[71] For instance, in Chapter 12[13], on the marvellous animals of India, the RB, in the passage beginning _Ac aniueil gỏyllt anhegar..._, adds _lle bo y tan_ and _o helir_, which are lacking in the WB text.[72] The Philadelphia text, which has _aniueil gỏyllt anhygar a phop peth a'r ỏrth ỏynepo idaỏ a'e korn y llad, ac ny ellir y doui..._, closely follows the WB reading: _aniueil gvyllt anhegar. A phob peth a vrthỏynnepo idaỏ ae gornn y llad. Ac ny ellir y douy._[73] Later in the same chapter, the Philadelphia and WB texts agree on _llasỏot_ against RB-B _llyssywot_; as they do on _wrach_ (WB) and _vrachiod_ (Philadelphia) against RB-B _wraged_. Finally, towards the end of the same chapter, the WB reading _Ac yn y mor hwnnỏ y maent malwot_ is followed verbatim by the Philadelphia manuscript, while the RB-B text reads _ac yn y mor hwnnw y magant_, omitting _malwot_.[74] The most illustrative example is, perhaps, Chapter 21[22] 'On Europe', where the Philadelphia and WB manuscripts share

---

[70] See p. 55 of the present edition.

[71] Falileyev, '_Delw y Byd_ Revisited', p. 73. The example of agreement cited by Falileyev of RB-B with Philadelphia against WB is an omission of a word in WB in Chapter 20[21], where the phrase rendered as _ac gỏedy goresgyn y rannỏys y ỏlat udunt_ (Philadelphia, f. 2r, lines 13–14) and _A gỏedy y goresgyn y rannỏys y ỏlat udunt_ (RB-B) according to Falileyev is 'slightly differently paraphrased in the White Book'. The paraphrase, however, is limited to the omission of _y rannỏys_ in the White Book, which reads: _Ac gỏedy goresgyn y wlat ydynt_ (WB, f. 2v, line 19). This omission does not preclude the possibility of a common exemplar for the Philadelphia and White Book texts. Falileyev's example of the agreement between the White Book and Philadelphia manuscripts against RB-B is, however, significant: in Chapter 19[20] they read _y mae dinas a oruc_, while the Red Book has _y mae ephesius dinas a wnaeth_; Falileyev, '_Delw y Byd_ Revisited', ibid. The text quoted throughout this discussion is from my transcription.

[72] Since this Version B chapter is not included in the current edition, for the WB text and RB-B variants, see _DB_, pp. 88–89.

[73] Philadelphia 8680.O, f. 1v, lines 1–16; WB, f. 2r, lines 2–3.

[74] Philadelphia 8680.O, f. 1v, lines 20–21; RB f.122v, col. 506, lines 34–35; WB f. 2r, line 6; for the WB and RB texts, see also _DB_, p. 89.

an addition in the form of a stray unidentifiable place-name at the end of the chapter, rendered variously *Chorsicca* (WB) and *Chosictu* (Philadelphia), and absent from the RB-B version.[75] In the same chapter, the WB and Philadelphia manuscripts both give the name of the Riphean mountains (as *Riphei mynyded* in WB and *Mynyd Ripheri* in Philadelphia), whereas the RB-B text omits the name, calling them *amryuael vynyded*. A third, though less significant variant from the same chapter can be given in support of this trend as cumulative evidence, for the RB-B derives the name of the Don from king *Tanaus* rather than *Tanais* as in WB and Philadelphia. We can therefore conclude that the scribe responsible for the Philadelphia text of *Delw y Byd* had access to an exemplar related to the WB.

The text of Rawlinson B 467 also undoubtedly belongs to Version B, and according to Falileyev it is sometimes closer to the RB and sometimes to the WB text.[76] The readings are indeed often very close, as demonstrated by Chapter 38[38], reproduced in the present edition.[77] However, in general, the text appears closer to the RB-B than the WB. For instance, in Chapter 40[40], relating to the tides (not reproduced in the present edition), the phrase *Kanys pan dyuo y lleuat y tyf* is present in full in both RB-B and Rawlinson, but the WB text omits *y lleuat*.[78] Similarly, in the same chapter the RB-B text (and, with minor spelling variation, Rawlinson B 467) reads: *mꝏyhaf vyd y llanꝟ rac <u>nesset vyd y lleuat. A phan vo hir y dyd y byd lleiaf y llanꝟ rac y bellet. Ac y myꝟn pob pedeir blyned ar bymthec y daꝟ y dechreu mal y lleuat</u>* (f. 124r, col. 513, lines 13–18). The underlined sections are omitted in the WB. In Chapter 41[41], on the ocean (not reproduced in this edition), where the RB-B reads *yn y dayar y mae eigaꝟn o'r dꝏfyr, a cherllaꝟ yr eigaꝟn y mae lleoed keu. A gogofeu llydan* (f. 124r, col. 513, lines 21–23) and Rawlinson B 467 reads *ac yn y daear y mae egyaꝟn or dꝏfyr a chyr llaꝟ yr eigyaꝟn honno y mae <?> keu <a?> gogofeu llydan* (f.71r, lines 13–16), the WB has *Ac yn y dayar y mae gordwyn a cher llaꝟ yr eigyaun honno y mae gogouev llydan a lleod keu* (f. 3v, lines 4–5), omitting the reference to the ocean and switching the order at the end of the phrase. Similarly, the final phrases of that chapter in the Rawlinson and RB-B texts agree against the WB: *Ac y myꝟn yr eigaꝟn trꝏy ogofeu y dayar* (RB-B, f. 124r, col. 513, lines 27–28) and *ac myꝟn yr eigyon trꝏy ogogofeu y dayar* (Rawlinson B 467, f. 72r, lines 4–5) versus *a ymgwyn trꝏy y gogouev y dayar. Ac eilꝟeith y crynnant o vaꝟr teruysc*[79] (WB, f. 3v, line 7). This precludes any suggestion that these extracts in the Rawlinson manuscript may derive from the WB. Meanwhile, the few instances of agreement between Rawlinson and the WB against the RB do not have equivalent significance.

---

[75] For the three variants of the entire chapter, see the present edition, p. 53.
[76] Falileyev, 'Delw y Byd Revisited', p. 73–74.
[77] See pp. 59–60 below.
[78] Here I quote the RB, f. 124, col. 513, lines 11–12.
[79] Corrected from *teruyst* in the manuscript.

Both have *honno*, omitted in the RB, in the passage quoted above from Chapter 41[41], and *a phan aner* to the RB's *am hanner* in the same chapter.[80]

Neither variant precludes the possibility that the Rawlinson manuscript may share an exemplar with the RB, although variation in the second of the two Rawlinson extracts, chapters 59[54] to 63[58], precludes the possibility that the text was copied from the RB itself.[81] Chapter 60[55] (on the winds) in particular provides some significant variants. This chapter is included in the present edition, with variants from all manuscripts, and is accompanied by diagrams to illustrate the variation.[82] It is worth noting, however, that the same chapter confirms the pattern of the Rawlinson B 467 fragments' agreement with the RB-B text against the WB, for the most significant variant is the omission in the WB of the four winds, due probably to eye-skip. These winds are present in the Rawlinson B 467 version.

Minor variants from other chapters (not reproduced in the present edition) include *Ac o hynny petwar prifwynt yssyt* in Chapter 59[54] 'On Wind', for *Ac or deudec hynny y mae pedwar prifwynt* (RB-B and, with minor orthographic variants, Rawlinson), and the transposition of *gulybôr a mellt* in the description of 'Auster and Nothus'. Minor variation between the RB-B and Rawlinson, such as the RB-B omission of *trayaneu* (WB *taranev*) in the description of Affricus only confirms that the RB-B and Rawlinson manuscripts are probably one exemplar removed from each other.

It therefore appears that the *Delw y Byd* extracts in the Rawlinson B 467 manuscript were taken from the exemplar of the RB-B text or at the very least a very closely related text. Meanwhile, the Philadelphia version of the text is closer to the WB version. The distance between the texts, however, seems to imply that more exemplars must have existed at some point, pointing to a manuscript tradition significantly more extensive than that which currently survives.

Structure

The Red Book and White Book Endings

As seen above (Table 1), the RB contains both versions of the text. Whilst deriving from different redactions of the *IM*, these two versions share a common feature that requires additional discussion. As noted by Falileyev,

---

[80] It is possible that the final sentence of Rawlinson Chapter 40[41] may have agreed originally with the WB.
[81] For the make-up of the Rawlinson extracts of *Delw y Byd*, see Table 1 above, pp. 8–9.
[82] For the chapter and diagrams, see below, pp. 62–65.

the continuous text of *Delw y Byd* in both the A and B versions in the RB, as in the Version B text of the WB, ends with Chapter 81[76] 'On Saturn' (reproduced in the present edition).[83] Thus, RB-A agrees with Version B against the Version A text of Peniarth 17 in its conclusion. (Peniarth 17 contains the only copy of the Welsh text which continues until the final chapter of the *IM* Book I.) The conclusion of the text at Chapter 81[76] complicates the issue of the relationship between the A and B versions of *Delw y Byd*. Indeed, both the WB (which represents Version B) and RB-A versions have a concluding 'here ends' phrase at the end of the relevant chapter, a phrase which is lacking in the RB-B version. This common feature would not be surprising for the RB-B and WB texts, both of which contain Version B of *Delw y Byd*, suggesting that that version was based on a single fragmentary text (whether Latin or Welsh), which ended at that chapter, but the conclusion of the RB-A text at the same chapter is unexpected. The situation is outlined in Table 2 below.

	Version A	*Version B*	
	Red Book A	Red Book B	White Book
'here ends' phrase after Chapter 81[76]	o	–	o
miscellaneous additional chapters following Chapter 81[76]	Bridging passage, 88[83]	74[69], 77[72], 67[62]–69[64]	–

Table 2: Endings and additional chapters in RB-A, RB-B, and WB

The question is whether this apparently common feature might be mere coincidence. Indeed, as can be seen from the table above, both RB texts contain additional miscellaneous material deriving from *Delw y Byd*, appended after Chapter 81[76]. As could be expected, given the different origins of the A and B versions of the text, the miscellaneous additional chapters which follow Chapter 81[76] in the RB-A version are not the same as those which follow it in the RB-B version. It must also be noted that while the continuous sequential part of the RB-B version ends at Chapter 81[76], this is neither acknowledged in the text by a concluding phrase, nor does it in fact terminate the text itself. The additional chapters in the RB-B text are, as mentioned above: 74[69], 77[72], and 67[62] to

[83] Falileyev, '*Delw y Byd* Revisited', p. 77. For Chapter 81[76], see p. 69 of the present edition.

69[64].[84] The RB-A text is also, despite featuring a concluding phrase at the end of Chapter 81[76], *A llyna diwed y llyuyr hvnn*, suggesting completion, followed by additional material, in this case corresponding to *Delw y Byd* Chapter 88[83] (on planetary distances; attested otherwise only in Peniarth 17), preceded by a bridging passage.[85] This is examined in more detail below.

The Red Book Additions

Neither the RB-A nor the RB-B text actually ends at Chapter 81[76], as additional miscellaneous chapters are present in both. In the RB-A version, the additional chapter is seen as a separate text, while in the RB-B version the additional chapters are seen as part of the same text. This is shown by both the presence of the 'here ends' phrase after Chapter 81[76] in RB-A (absent in RB-B) and by the layout and decoration, which mark the addition as a separate text in RB-A but not in RB-B. It is unclear at what point these additional chapters might have been added to either version.

As already noted, additions to RB-A consist of Chapter 88[83], preceded by a bridging passage which does not appear to have an equivalent anywhere in the *IM* or the other texts of *Delw y Byd*. The chapter deals with planetary distances but displays a number of peculiarities. Firstly, in the RB-A addition, Venus is omitted from the enumeration of the planets (possibly due to eye-skip when copying). Secondly, the names given to the planets are in a form not attested elsewhere among our *Delw y Byd* texts: *mercuriόm, marten, Jouen*, and *saturniόm*. These correspond to the original Latin forms (including case-endings) in the passage.[86] Whilst instances of case-ending retention can be found in both A and B versions of *Delw y Byd*, a greater tendency to maintain the Latin endings in the grammatical form in which they occur in the original text is generally a feature of the B version (both in the WB and in the RB-B text), primarily in place-names, as in *ac o garyat* Caonis *y vrawt y gelwis* Caoniam (WB 25[27]), rendering the Latin *et ob amorem fratris sui* Chaonis Chaoniam *nominavit*.[87] RB-A renders the names as *Caon* and *Caonia*.[88]

Returning to the planet names in Chapter 88[83], Peniarth 17, which contains the equivalent passage, changes these forms to *Mercurius, Mars, Iupiter*, and *Sadurn*.[89] Furthermore, in the RB-A addition, *lloer* is used instead of *lleuat*, the

---

[84] See p. 8 above.

[85] Both reproduced in the present edition, pp. 69, 71.

[86] Cf. *IM*, ed. by Flint, p. 80: *Mercurium, Martem, Iovem, Saturnum*.

[87] *IM*, ed. by Flint, p. 60.

[88] For the RB-A text, see *DB*, p. 41. Other instances include the WB's use of *Ierusalem* and *Sarraceni* for RB-A *Kaerusalem* and *Sarassinyeit* in chapters 15[16] and 16[17]; see *DB*, pp. 33, 35, and 89–90. For further examples, see also, for instance, Chapter 21[22] in the present edition, p. 53.

[89] Both variants of the passage are reproduced in the present edition in full, pp. 70–71.

form used in the equivalent chapter of Peniarth 17 and also with reference to the moon elsewhere in the *Delw y Byd* texts. The form *lloer* is otherwise unattested in *Delw y Byd*.[90] The addition in RB-A is also at variance with Peniarth 17 and the Latin original in terms of distances given, as shown in Table 3 below.

	IM[91]	RB-A addition	Peniarth 17
Earth to Moon	15,625	*15,635*	*15,620*
Moon to Mercury	7,812.5	7,812.5	7,812.5
Mercury to Venus	'same'	*N/A* (Venus omitted)	'same'
Venus to Sun	23,436	(Mercury to Sun) 23,436	*24,436*
Sun to Mars	15,625	15,625	*12,622*
Mars to Jupiter	7,812.5	7,812.5	*7,812*
Jupiter to Saturn	'same'	7,812.5	'same'
Saturn to firmament	23,437	*23,436*	*24,336*
Earth to firmament	109,375	*109,335*	*109,345*
[actual total]	[109,373]	[101,569.5]	[108,263]

Table 3: Planetary distances in *IM*, RB-A, and Pen.17, Chapter 88[83]

Where the number given in the Welsh versions differs from the original, it is in bold italics. 'Same' is written where the number is not given in the text but is designated as being the same as the previous. It will be noted that on the whole the RB-A addition is much closer to the original text, with fewer errors, than is the Peniarth 17 text.

It does not appear at present possible to assign this additional Chapter 88[83] of the RB to either of the known versions of *Delw y Byd*. Since the equivalent chapter does not survive for Version B (and indeed, given the possibility that Version B originally ended at the Saturn chapter, 81[76], as suggested by the concluding phrase in the WB, it is possible it may never have had an equivalent chapter), it is difficult to suggest that the differences noted above are due to the addition in question belonging to the B rather than the A version of *Delw y Byd*. There appear to be three possibilities. The first is that this additional chapter reflects an earlier stage in the transmission of Version A, wherein the Latinate

While the WB text does not have the equivalent passage, in earlier chapters it uses the following forms: *Mercurius, Mars,* and *Iouis* (it omits Saturn).

[90] Elsewhere in *Delw y Byd* the only instance is in the RB A-516 Fragment Chapter 74[69] in the phrase *aroyd sychhin ar y llaonlloer*.

[91] *IM*, ed. by Flint, p. 80, taking into account variants represented by the group of manuscripts to which Cambridge, Corpus Christi College MS 66 belongs.

endings for the planets were retained, of which it is the only surviving fragment. The second is that it belongs to a lost post-Chapter 81[76] section of Version B.[92] The third possibility is that this chapter represents an otherwise unattested third version of *Delw y Byd*, the exact relation of which to Versions A and B is as yet impossible to determine. Yet again, we appear to be faced with the remains of a more extensive manuscript tradition.

The RB A-516 Fragment

The additions at the end of the RB-B (column 516) are a different phenomenon. Here, we are faced with what is in essence a short fragment of *Delw y Byd*, comprised of chapters 74[69], 77[72], and 67[62] to 69[64].[93] The text of the fragment corresponds closely to the RB-A version of *Delw y Byd*. It is therefore henceforth referred to as the 'RB A-516 Fragment', since this text reflects Version A but occurs at column 516. The source of these passages and the motivation for their addition to the text is unclear. At least for 74[69] (Moon) and 77[72] (Sun), the section of the chapters reproduced corresponds to a section not present at all in Version B. This may explain the addition of this extra material at this point in the manuscript. It is also possible that the chapters were already present in Hywel Fychan's exemplar, or that he may have had access to extracts which he recognized as relating to the same text. That extracts of *Delw y Byd* did circulate separately is attested by Rawlinson B 467. The idea of extracts circulating separately and in a different manuscript context would explain the transposition in the order of the chapters, as well as the omissions.

Comparing the RB A-516 Fragment to Version A of *Delw y Byd* we find that the correspondence is much closer to the RB-A text than it is to Peniarth 17, with variation limited largely to orthography.[94]

The presence of two versions in the RB, along with additional material derived at least partially from a possible additional version, is not only evidence

---

[92] Whilst we have no evidence for the existence of this section of the B text, it would be inadvisable to argue for its non-existence on the basis of the concluding 'here ends' formula in the WB. We have a similar formula in the RB-A text, but we also have the rest of the A text surviving in Peniarth 17. The formula therefore means very little as regards the existence or otherwise of more text. Cf. also the use of the same formula in the short version of *Historia Peredur fab Efrawc*; see N. I. Petrovskaia, 'Dating *Peredur*. New Light on Old Problems', *Proceedings of the Harvard Celtic Colloquium*, 29 (2009), 223–43 (pp. 229–30).

[93] Lewis and Diverres print these separately under 'Atodiad' in their edition, *DB*, p. 112, and Falileyev comments: 'The order of these chapters is peculiar 89 [*sic*], 72, 62, 63 (Flint's 74, 77, 67, 68), and these are not attributed to any of the extant versions of Welsh translations. A similar order of chapters has not been so far identified in the versions of the original text'; '*Delw y Byd* Revisited', p. 74.

[94] See pp. 66–68 of the present edition, and discussion on p. 41. It can be seen that the punctuation and capitalisation also match in the two texts. As a comparison of the RB-A and RB A-516 texts shows, the variation is minimal, showing that they are closely related.

for the circulation of multiple versions of the text and for their interest, but is also an example of collation, which we find attested elsewhere in the manuscript tradition of the Welsh Charlemagne material and the Welsh *Transitus Mariae*.

The Peniarth 17 Text

In contrast to the complex construction of the RB fragments, the text preserved in Peniarth 17, designated the best of the surviving witnesses by Lewis and Diverres, appears to be a straightforward copy of a single version, without interpolations.[95] The text, as we have it, is acephalous, beginning at the top of page 17 with the words *pan gerdo y duuyr*, and follows an incomplete text of the *Historia Gruffudd ap Cynan*.[96] The end of *Historia Gruffudd* is missing from a section corresponding to page 23 line 23 onwards in the edition.[97] Page 17 of Peniarth 17, at the top of which begins the acephalous text of *Delw y Byd*, is the beginning of a full quire, and we are therefore missing a section of the text anywhere in size from half a page to a full quire.[98] The remainder of the text of *Historia Gruffudd ap Cynan* (if the text corresponded roughly to versions preserved in other manuscripts) is approximately 2100 to 2200 words, and the pages of Peniarth 17 carry approximately 370 to 380 words per page.[99] Thus, there does not seem to be enough of the *Historia* missing to have filled a quire,

---

[95] Peniarth 17, probably of Aberconwy origin, shares its scribe with Peniarth 14 (1–44) and the Book of Aneirin (Cardiff 2.81); *MWM*, pp. 58, 75. For more on Peniarth 14 (1–44), see Ingo Mittendorf, 'Y Groglith *Dyw Sul y Blodeu*: Die mittelkymrische Passion nach Matthäus in Peniarth 14 und Havod 23', in *Übersetzung, Adaptation und Akkulturation im insularen Mittelalter*, ed. by Erich Poppe and Hildegard L. C. Tristram (Münster: Nodus Publikationen, 1999), pp. 259–88. Note, however, that the orthography and language of the Book of Aneirin differ significantly from both Peniarth 14 (1–44) and Peniarth 17; Mittendorf, 'Y Groglith', p. 263 n. 10. This is discussed in greater detail below, in the 'Notes on the Commentary' section. For a discussion of the script of Peniarth 17, see also *Historia*, ed. by Evans, pp. cclii–cclv.

[96] For a transcription of this manuscript, see G. R. Isaac and S, Rodway, 'Peniarth 17', *Rhyddiaith Gymraeg o Lawysgrifau'r 13eg Ganrif: Testun Cyflawn* (2002), now available in an updated edition: G. R. Isaac, Simon Rodway, Silva Nurmio, Kit Kapphahn, and Patrick Sims-Williams, *Rhyddiaith Gymraeg o Lawysgrifau'r 13eg Ganrif: Fersiwn 2.0* (2013), available at <https://cadair.aber.ac.uk/dspace/handle/2160/11163> [accessed 18 June 2019]. Note that the contents of the CADAIR Open Access Repository are due to be moved to PURE, where the database will be findable under the keywords 'Welsh & Celtic Studies' and 'Rhyddiaith'. I am grateful to Simon Rodway for this information.

[97] *Vita Griffini Filii Conani: The Medieval Latin Life of Gruffudd ap Cynan*, ed. by Paul Russell (Cardiff: University of Wales Press, 2006), p. 3 n. 11; *Historia*, ed. by Evans, p. 23.

[98] Judging from the state of the recto of the first folio of the quire, it seems a distinct possibility that the quire had lost its companion at an early stage. Page 17 has sustained heavier damage than all the other pages in the quire and is considerably darker than pages 18 or 19. The preceding pages, bearing the *Historia Gruffudd vab Kenan*, are similarly affected, but to a much lesser extent. They are far more readable.

[99] Calculations based on the text as presented in Evans' edition.

as there is enough for only approximately five and a half pages. Since the end of *Delw y Byd* in Peniarth 17 corresponds to the end of the text of Book I of the *IM*, it is not beyond reason to assume that the manuscript may originally have contained the whole of Book I.[100]

Based on the word count of the corresponding section in the RB, if Peniarth 17 originally contained a complete text of *Delw y Byd*, then approximately 4800 words are missing from its beginning. At *c.* 400 words per page, as observed in the *Delw y Byd* section of the manuscript, the result is approximately twelve pages.[101] The total loss, based on the two texts, would approximate seventeen to eighteen pages, which would be just one or two pages more than an eight-folio quire. Given that the calculations are approximate and based on the amount of text attested in other manuscripts, it is possible that one quire, containing the end of the *Historia Gruffudd ap Cynan* and the beginning of *Delw y Byd*, is all that is missing from this manuscript.[102] Considering the size of the surviving manuscript, however, the loss might have been considerably greater.[103]

The Philadelphia and Rawlinson Fragments

The question of loss brings us to the subject of the fragments preserved in the Philadelphia and Rawlinson manuscripts. The Philadelphia manuscript contains two fragments of *Delw y Byd*, corresponding to chapters 11[12] to 12[13] on f. 1 and 18[19] to 24[26] on f. 2. As observed by Ben Guy, the lacuna between the two fragments is probably due to the loss of a bifolium.[104] Whilst the manuscript as a whole is best known, perhaps, for its association with Hywel Fychan, the copy of *Delw y Byd* preserved in it originated in a different context and bears no relation to the section written by that scribe. The text appears to date from the fourteenth or fifteenth centuries, and the scribal hand shares some similarities with that of Hywel Fychan, prompting Guy to suggest that these scribes were

---

[100] The *Delw y Byd* text ends at page 26 at the words *brenhin // er englynion*. The words *er englynion* are added at the bottom of the page, preceded by a paragraph mark. The phrase *arwyd temestyl uyd* at the top of page 27, above the beginning of *Diarhebion*, has led Lewis to remark on the possible loss of more text; *DB*, p. xiii. It is however probably a catch phrase to identify which gathering goes with which, given that page 27 is the first in a new gathering. Indeed, the exact phrase occurs at lines 1–2 of page 26.

[101] Note that the difference in the word count per page might be due to the different conventions of word separations employed by Evans's edition and the present edition. However, within each text, the calculations were consistent. The difference in the words-per-page count between texts should not matter for the ultimate quire count for each text.

[102] This supports the conclusion reached by Evans; see *Historia*, ed. by Evans, p. ccli.

[103] Peniarth 17 consists of only two quires of eight folios each. The first quire is occupied by the *Historia Gruffudd vab Kenan*, the second is shared by *Delw y Byd* and the *Diarhebion* ('proverbs').

[104] Guy, 'A Welsh Manuscript in America', p. 99 n. 14.

contemporaries.[105] However, leaf sizes and writing space suggest that the only original companion of the text now preserved in the same manuscript is the Calendar written in a hand designated by Guy as 'C' (ff. 3–4), dated by him to the same period as the *Delw y Byd* scribe.[106] According to Guy, these two sets of folios were probably added to the manuscript in the second half of the sixteenth century, possibly in South-East Wales.[107] It is also in that area that scribe 'C' appears to have been active.[108] Since the *Delw y Byd* bifolium, albeit written by a different scribe, seems to have originally formed part of the same manuscript as that of scribe 'C', one could put forward the tentative suggestion that the *Delw y Byd* fragment of Philadelphia 8680.O originated in South Wales.

Rawlinson B 467 stands out from the rest of our manuscripts for several reasons. In the first place it is a thematic compilation unlike any of the others: a collection of medical texts.[109] In the second place, it contains extracts from our text, selected purposefully for their content, rather than fragments, preserved through chance. The manuscript is also in its original binding, facilitating the identification of the original context of our text in this manuscript.[110] The manuscript appears to be uniform not only in its scientific focus but also in date: its four scribal hands all date from the late fourteenth or the early fifteenth centuries.[111] The extracts from *Delw y Byd* are the last item in Hand C, responsible for a number of texts in the same manuscript, including *Deuddeng Arwydd* and *Llythyr Aristotlys at Alecsander: Pryd a Gwedd Dynion*, the Welsh translation of part of the Letter of Aristotle to Alexander, better known as *Secretum Secretorum*.[112] The manuscript contains chapters 38[38] to 40[40, 41]

---

[105] Guy, 'A Welsh Manuscript in America', p. 102 and ibid. n. 26.

[106] Guy, 'A Welsh Manuscript in America', p. 102. According to Guy, the inclusion of St Kenlem of Winchcombe (Gloucestershire) and Thomas de Cantilupe (Hereford) indicates a South-Eastern Welsh provenance for the calendar; ibid., p. 107.

[107] Guy, 'A Welsh Manuscript in America', pp. 107–08

[108] Guy, 'A Welsh Manuscript in America', p. 107.

[109] For a brief introduction to medieval Welsh medical manuscripts, see Morfydd E. Owen, 'Meddygon Myddfai: A Preliminary Survey of Some Medieval Medical Writing in Welsh', *Studia Celtica*, 10–11 (1975–1976), 210–33.

[110] *MWM*, p. 43.

[111] *RG*, TEI header for Rawlinson B 467 [accessed 1 March 2016].

[112] Not to be confused with the 'Letter of Alexander to Aristotle about India'; for more on the *Secretum Secretorum*, see *Secretum Secretorum: Nine English Versions*, ed. by M. A. Manzalaoui (Oxford: Oxford University Press, 1977); Manzalaoui, 'The *Secreta Secretorum* in English Thought and Literature from the Fourteenth Century to the Seventeenth Century with a Preliminary Survey of the Origins of the *Secreta*' (unpublished DPhil dissertation, University of Oxford, 1954); Stephen J. Williams, *The Secret of Secrets: The Scholarly Career of a Pseudo-Aristotelian Text in the Latin Middle Ages* (Ann Arbor: University of Michigan Press, 2003). For an edition and translation of the Welsh text (from a different manuscript), see Ida B. Jones, 'Hafod 16', *ÉC*, 7 (1955), 46–75 (pp. 64–73). For a discussion of the text in a Welsh context, see Mark Williams, *Fiery Shapes: Celestial Portents and Astrology in Ireland and Wales, 700–1700* (Oxford: Oxford University Press), pp. 114–15.

and 59[54] to 63[58] of *Delw y Byd*, concerning water and the winds. It is worth noting that other medical manuscripts containing collections related to that found in Rawlinson B 467, such as Cardiff, MS 3.242 (Hafod 16), do not contain the extracts from *Delw y Byd*. It therefore appears that the inclusion of this text in the compilation was the initiative of the compilers of the Rawlinson manuscript or of the compilers of an earlier version in that tradition. The extracts fit thematically with the accompanying texts. They are preceded by a text identified in *RG* as *Wyth Rhan Pob Dyn*, 'The Eight Parts of a Man', which identifies the eight parts of man with earth, sea, sun, wind, clouds, stones, holy spirit, and the moon.[113] The text following the *Delw y Byd* extracts is a translation of a tract attributed to Galen and Hippocrates.[114] The extracts thus appear to have been brought into the collection to enrich its encyclopedic and informational value. Their use here is scientific rather than religious or antiquarian. Their inclusion in the medical collection has several implications for our understanding of the text's place in medieval Welsh thought. In the first place, it appears that it was perceived as a general encyclopedia from which one could extract information of relevance to different subjects. It also shows that medieval Welsh scholars actively sought additional information and additional sources for inclusion in their compendia.

Notes on the Editorial Method

General Observations

The existence of two separate surviving translations and their fragmentary nature preclude the possibility of publishing a complete text based on either a single manuscript or on the principles of a critical edition. Whilst for research purposes a complete edition of the surviving versions, including all of the manuscripts and fragments, would be ideal, the limitations on the size of the present edition, and its primary function as a text for classroom use, preclude such an approach. Given the availability of a previous edition presenting the complete texts of RB-A (with a small omission), Peniarth 17, and WB, and the accessibility of both transcriptions and manuscript images for the rest, it seemed appropriate to contribute to the available material by presenting a composite edition based on the structure of the underlying original text. As far as the text of individual chapters is concerned, this is a diplomatic edition, rendering the text as it appears in the manuscripts (with emendations made only in cases of

---

[113] For the text and translation, see Ida B. Jones, 'Hafod 16 (A Medieval Welsh Medical Treatise) (suite et fin)', *ÉC*, 8 (1959), 346–93 (pp. 382–85).

[114] Note, however, that this text belongs to a new unit within the manuscript, copied by a different scribe; Luft, 'Ansoddau'r Trwnc', p. 60 n. 28.

obvious errors, and noted in footnotes in all cases).[115] However, in order to bring to the fore the structure of the text, section divisions are introduced and the text is treated as a continuous whole, despite the fact that no one manuscript of *Delw y Byd* preserves all of it. The only violence made to the original structure of the text is in the removal into the appendix of the two 'letters' which precede Chapter 1[1] in the RB-A text and serve as an introduction or preface to the text. Both the Welsh and the original Latin in the letters present difficulties of interpretation which are not typical of the rest of the text.

The guiding principle in the present composition is, therefore, to provide as clear an idea as possible of the information *Delw y Byd* made available to its medieval Welsh audiences, as well as the structured way in which this information was presented. The latter is deemed particularly important, as any diplomatic edition, or even a composite edition based primarily on manuscript structures, would invariably push the original organising principles of the work to the background. Since Version A of *Delw y Byd* is the only text surviving fully (albeit across two manuscripts), it will be used to provide the baseline of the edition. Variant chapters from other manuscripts, from Version B, will be provided where the variation is of particular interest.

Any selection is a subjective undertaking. The selection of chapters for inclusion here has been made on the basis of three considerations: structure, use, and representativeness. In the first place, included are chapters crucial to understanding the structure and composition of the text (and the structure of the world which it presents). These include chapters which occur at the ends and beginnings of the major sections into which the text is separated, even when the information contained in the chapters themselves is of limited interest. In the second place, the edition includes chapters which present material of particular interest. One of the guiding principles of selection here was that chapters are preferred which have been, either in the original Latin version or in the Welsh, the subject of study and discussion. Finally, chapters which are representative of the particular types of information presented are included. *Delw y Byd* often has sequences of chapters dealing with particular sequences of objects (e.g. countries, planets, constellations). In these cases only one or two from the sequence are reproduced. It is hoped that this selection will give the reader as complete an idea as possible about the encyclopedia, its organising principles, its subject matter, and the origins of the information it contains. A full list of *Delw y Byd* chapters is provided below (following the introduction and preceding the text), identifying chapters and variants reproduced in the present edition.

As mentioned above, I have divided the text into sections, corresponding to the structure the text outlines in its early chapters. Whilst these sections are

---

[115] For a detailed discussion, see pp. 28, 31–32, and 38–41 below.

made clear in the texts themselves, titles are not provided in the manuscripts, and the section titles provided in the present edition are mine.[116] Explanatory passages have been added as commentary to provide an overview of those sections of text which have been omitted from the present edition.

Since the objective of this edition is to give the reader a glimpse of the material as it was available to medieval audiences, editorial intervention has been kept to a minimum. Text deleted in the manuscript either by strike-through lines or by employing a *punctus delens* has been reproduced with strike-through to denote its deletion by the scribe. Text added by the scribe as superscript correction has been retained as superscript. Expanded abbreviations are given in italics. Intervention and emendation have been kept to a minimum, and are always commented on. In the very few cases where an error in the manuscript is apparent, I have corrected the reading (e.g. *heul* where the manuscript reads *heu*) and given the original reading in the footnote. Where the text is illegible, I have marked the illegible words as [?]. Supplied text is given in square brackets.

Some cosmetic alterations have been made to the text to facilitate use, as it was deemed unnecessary to replicate the provision of transcriptions available on *RG*.[117] Certain changes, such as the introduction of chapter divisions and the introduction of additional capitalisation, were deemed necessary in order to make the text more accessible to the student reader (who might appreciate the distinction of personal and place-names that capitalisation affords) on the one hand, and more useful to the researcher (who may wish to compare the text to the original Latin, for instance) on the other. Punctuation marks and capitalisation are employed in the manuscripts with a great degree of consistency, and I have retained these patterns in order to maintain the structures which they create, even where they appear counter-intuitive, so long as understanding is not impeded thereby. The changes made are described below in greater detail.

The principle guiding both the changes and the commentary is that recently outlined by Rudolf G. Wagner for Chinese studies: 'the shared assumption is that texts intend to make a sense that can be understood, and that it is the duty of the commentator to pave the way for the reader's understanding by smoothing out the obstacles on the character surface of the text'.[118] Whilst the approach of assigning meaning to texts that they do not necessarily have is

---

[116] The only exception is the section on air, p. 62.

[117] *RG* [accessed 23 October 2015]. The transcriptions of all versions of *Delw y Byd* apart from Peniarth 17 are available from this excellent online resource. Isaac and Rodway, 'Peniarth 17', is a transcription of Peniarth 17.

[118] Rudolf G. Wagner, 'Does This Make Sense? Reading Sinological Translations', in *Zurück zur Freude. Studien zur chinesischen Literatur und Lebenswelt und ihrer Rezeption in Ost und West*, ed. by Marc Hermann, Christian Schwermann, and Jari Grosse-Ruyken (Sankt Augustin: Institut Monumenta Serica, 2007), pp. 767–76 (p. 771).

highly problematic, it seems justified when reading a scientific or encyclopedic text to expect that it is supposed to carry meaning and to be intelligible. In the case of *Delw y Byd* the process of interpretation is aided by the availability of the original from which the text was translated and against which difficult passages can be checked. Thus in the present edition explanatory notes are provided for passages which appear to contain errors introduced by the translators or scribes, as well as for those passages whose significance may not be immediately obvious.

In *Delw y Byd*, as in other medieval texts, there are what C. S. Lewis called 'treacherous passages', sections of text 'which will not send us to the notes. They look easy and aren't'.[119] Such passages, along with those that are 'manifestly hard', have received heavy annotation. The choice made in this edition has been to maintain to some degree a brevity in the introduction and to provide notes for the instruction of readers, an option made possible by the format of the enterprise. The notes, nevertheless, are placed at the end of the volume, so that the reader may consult them at leisure and so that they may not prevent a smooth reading of the text.

Chapter Numbering

The chapter numbering for *Delw y Byd* has been complicated by the existence of two different systems of chapter division. The first system was used by J.-P. Migne in his Patrologia Latina edition, which essentially was a reprint of the 1677 printing of the edition originally published by André Schott (1552–1629).[120] This system was followed by Lewis and Diverres in *DB*. The chapter divisions were changed by Valerie Flint in her 1982 edition of the *IM*. In the present edition I follow the chapter numbering of Flint's edition and provide the corresponding Migne/Lewis and Diverres chapter numbers in square brackets.[121] For example, 32[33] means the chapter number is 32 in Flint's edition and 33 in both Migne and *DB*.[122]

There is a pattern of capitalisation and some (though not many) coloured initials present in all of the fragments. Whilst these seem to indicate a sense of chapter divisions, the chapter division system followed in the edition corresponds, for the sake of ease of use, with that already established in

[119] Lewis, *The Discarded Image*, p. ix.
[120] The text edited by Schott printed in the *Maxima Bibliotheca Veterum Patrum*, vol. 20 (Lyon: 1677), fol. 963 onwards, referred to by Flint (*IM*, ed. by Flint, p. 44 n. 1), is a reprint post-dating Schott's death. Earlier printed editions of the *IM* include that published at Speyer in 1583 by Bernhardus Albinus, with an introduction by Johannes Trithemius (1462–1516).
[121] Petrovskaia, '*Delw y Byd*' and 'La disparition du *quasi*'.
[122] Flint provides chapter titles, based on headings found in some of the manuscripts; I have omitted these.

previous editions for *Imago mundi/Delw y Byd*. In cases where the Welsh text does not seem to follow the chapter divisions conventionally given to this text, I have reproduced the text as a continuous paragraph, giving the chapter number as e.g. <35[36]> within the text. The angle brackets are maintained in chapter references in the glossary, indices and notes for ease of reference.

Capitalisation and Punctuation

A system of capitalisation and punctuation is present to some degree in all of the manuscripts used. Punctuation is lacking only in the Rawlinson manuscript, but capitals are used. The only type of punctuation present in the manuscripts is the *punctus*. It is employed for pauses, quotations, to introduce important names, and to separate out roman numerals. This corresponds to common medieval practice but requires adjustment in order for the text to become accessible to the modern reader.[123] As a general principle, the original capitalisation/coloured initial pattern of the manuscripts is followed in establishing sentence divisions. Further alterations to the punctuation, and the principle for the application of these alterations, are described in detail below.

The treatment of punctuation in any edition requires a degree of explanation and commentary.[124] Medieval punctuation differs from modern conventions, and it has been customary for editors of medieval texts to summarily modernise punctuation even when the text is not otherwise emended.[125] There has been a recent trend in the analysis of medieval manuscripts and texts, however, involving a focus on the logic behind the use of punctuation, and demonstrating that this use is not arbitrary. Punctuation is a crucial part of the medieval text, particularly the material text as preserved by the manuscript, and is an aid to understanding the structure of the text and as well as medieval reading practices.[126] The function of modern punctuation is grammatical, whereas medieval punctuation is rhetorical in nature, reflecting the difference in reading practice (silent vs voiced).[127] The resulting suggestion is that it might be worth treating punctuation in the same way as orthography is treated, keeping the text as close as possible to the reading of the manuscript, with emendation only

---

[123] M. B. Parkes, *Pause and Effect: An Introduction to the History of Punctuation in the West* (Aldershot: Ashgate, 1992), p. 42.

[124] Often omitted, as Mary-Jo Arn points out; 'On Punctuating Medieval Texts', *Text*, 7 (1994), 161–74 (p. 172).

[125] See, for instance, Simon Horobin and Jeremy Smith, *An Introduction to Middle English* (Edinburgh: Edinburgh University Press, 2002), p. 20. Although based on analysis of medieval English texts, their observations are valid for medieval texts generally.

[126] Arn, 'On Punctuating Medieval Texts', pp. 162–63.

[127] Raymond Clemens and Timothy Graham, *Introduction to Manuscript Studies* (Ithaca, NY: Cornell University Press, 2007), p. 82.

where absolutely necessary for understanding.[128] The objective of this edition is to present the reader with a glimpse into the geographical and astronomical information found in *Delw y Byd* as it was available to be understood (and misunderstood) by medieval audiences. The objective therefore is to keep as closely as possible to the medieval text, punctuation included. As Arn points out, whilst the modern reader is unaccustomed to ambiguity, his or her medieval counterpart 'must have accepted the need to stop and decide between two or more possible readings'.[129] In a modern environment, the text intended for translation and analysis in the classroom represents the best possible equivalent for this medieval mode of reading and one of the few cases where following 'different courses in different readings of the same text' is possible and even recommended for the modern reader.[130] I have therefore chosen to maintain punctuation as it is present in the manuscripts insofar as was possible.

In some cases, the distinction between different types of punctuation is already present in the manuscripts, which appear to differentiate between types of stop through the use of capitalization and red coloured initials. The RB-A (and to some degree RB-B) texts and the Peniarth 17 text occasionally combine the use of *punctus* with capitals and coloured initials. The variation in these combinations is not random and I have chosen to maintain the distinctions, rendering them through the means of modern punctuation marks (comma and full stop). In order to provide some transparency to this editorial process and preserve as much as possible the original visual nature of the text, I have used **bold** type to indicate the use of red in the manuscript throughout the edition. Although capitalisation and colouring do not always correspond in the manuscripts, they are for the most part used to distinguish sentence units.[131] Thus, as a general rule, where a capital or colour was used by the scribe to mark the beginning of a new sense unit, the preceding punctuation mark has been rendered as a full stop. Where no such distinction is made, a comma is given.

---

[128] For a discussion and overview of recent trends, see, for instance, Javier Calle Martín, 'Punctuation Practice in a 15th-century Arithmetical Treatise (MS. Bodley 790)', *Neuphilologische Mitteilungen*, 105 (204), 407–22 (pp. 407–08). See also Arn, 'On Punctuating Medieval Texts', p. 162.

[129] Arn, 'On Punctuating Medieval Texts', p. 165.

[130] Arn, 'On Punctuating Medieval Texts', p. 165. Arn also argues that in texts intended for students the editor should take on the teacher's role by guiding the student through the text by means of punctuation, since the student would have sufficient problems with meaning and context; ibid., pp. 168–69. I believe that in the case of *Delw y Byd* such guidance can be provided in exceptional cases where medieval punctuation hinders understanding, but given that such cases are sufficiently rare in the text, they cannot be used to justify systematic editorial intervention.

[131] There are some exceptions to this. In some cases, capitalisation is present in lists, as in the list of the elements in Chapter 3[3], where the coloured initials mark elements. It is probable that these are marked because the names of the elements correspond to sections of the text.

Notable exceptions to this rule include punctuation marks preceding lists, which have been rendered as colons.

The Philadelphia text makes very limited use of the *punctus* and capitalization, and thus punctuation has simply been provided throughout for all examples from that manuscript, for ease of reading. The WB text consistently has capitals following the *punctus*, and thus where a comma was necessary for clarity (in my interpretation of the text) I have introduced these and provided an interpretative commentary where necessary. In all cases where punctuation is added, it is underlined in the text to distinguish it from punctuation already present in the manuscripts.

Additional capitalisation has been introduced for place-names and personal names for the readers' convenience. For the sake of consistency and clarity, both parts of place-names, such as 'Mor Mawr', 'Mor Ynys Pont', or 'Mynyd Ethna', have been capitalised where the name follows the noun.

In cases where punctuation marks appear on both sides of a place-name or title they are treated as quotation marks. For ease of understanding, quotation marks have also been provided when the text discusses or explains a term, particularly after introductory phrases using *a elwir* or *dywedir*. The term in question is put in quotation marks (often these correspond to original punctuation marks in the manuscript). Most of the manuscripts appear to have a very consistent scheme which yields well to this treatment. Thus, changes made through other considerations are few, and these are noted as they occur. Sometimes a *punctus* appears to play the double function of quotation mark and comma, and in these cases both are introduced. I have aimed not to remove original punctuation marks unless absolutely necessary. A note is made wherever punctuation has been removed.

Language and Orthography

As a translation, *Delw y Byd* displays many of the linguistic characteristics observed in medieval Welsh translated texts, including constructions such as *yr hwn/yr honn*.[132] The discussion below provides some examples of the problems faced by the translators and the linguistic choices that they made. Although a full linguistic study of *Delw y Byd* is impractical here, some general observations can be made, in particular contrasting it with some of the better-studied translated narrative texts.[133]

---

[132] See John Morris-Jones, *Welsh Syntax: An Unfinished Draft* (Cardiff: University of Wales Press, 1931), p. 104 (ii) Note; and *GMW*, p. 66 n. 2, and discussion below, pp. 34–36. For a brief discussion of 'translator's style', see *Cyfranc Lludd a Llefelys*, ed. by Brynley F. Roberts (Dublin: Dublin Institute for Advanced Studies, 1975), pp. xxviii–xxxii.

[133] These narrative texts broadly fall into two categories: historical texts, such as the chronicles, known as the *Brutiau*, and narrative texts proper, such as the Charlemagne

There is some difference between the approaches to translation exhibited by the translators of *Delw y Byd* and those of the narrative texts. The factors influencing these differences correspond to those suggested by Brynley Roberts: differences in the purpose of the text and its target audiences.[134] In contrast to the narrative text of *Can Rolant* analysed by Luciana Cordo Russo, regarding which her conclusions were that the translator, in the course of compromising between the original and the target language 'created a sort of individual style', the translator(s) of *Delw y Byd* appear to have opted for maximum proximity to the original.[135] This appears to reflect a difference in genre, since Erich Poppe and Regine Reck's analysis of *Ystorya Bown o Hamtwn* shows a tendency to adapt the text to medieval Welsh narrative conventions, with 'accommodation of the foreign plot to Welsh literary conventions and the expectations of the new audience'.[136] The translator of *Bown* appears to have only followed his source in the case of structural organisation (explicit authorial marking of scene or protagonist transition) and particular exclamatory or formulaic phrases that have no native Welsh equivalent, but not in the case of syntactic structures.[137] As Poppe and Reck note,

> the Welsh redactor implemented major changes on the levels of narrative structure and style. They are intended to accommodate the foreign plot to Welsh literary conventions and to the expectations of its new audience, and can be detected on all levels of the text, from the macro-form and its syntactic organisation down to the micro-level of the use of specific tenses and idiomatic phrases.[138]

They describe instances in which the translator adopts a feature of the original as 'lapses'. In the case of the redactors of the two versions of *Delw y Byd*, the balance of the influence of the original versus the native conventions is reversed. It is possible that the transformation of syntactic features into native

material and the *Ystorya Bown o Hamtwn*. For further studies of the language of translated texts, see, for instance, Lloyd-Morgan, 'French Texts, Welsh Translators' and 'Rhai Angweddau ar Gyfieithu yng Nghymru yn yr Oesoedd Canol', *Ysgrifau Beirniadol*, 13 (1985), 134–45; Diana Luft, '*Awdur neu Dyallwr Ystoriau*: Theori a Chyfieithiadau Cymraeg yr Oesoedd Canol', *Llenyddiaeth mewn Theori*, 1 (2006), 15–40 and 'Tracking *ôl cyfieithu*: Medieval Welsh Translation in Criticism and Scholarship', *Translations Studies*, 9 (2016), 168–82; *Brut y Brenhinedd. Llanstephan MS. 1 Version*, ed. by Brynley F. Roberts (Dublin: Dublin Institute for Advanced Studies, 1971), pp. xliii–lx; Erich Poppe and R. Reck, 'A French Romance in Wales: *Ystorya Bown o Hamtwn*. Processes of Medieval Translations', *Zeitschrift für celtische Philologie*, 56 (2008), 129–64; L. C. Russo, 'Translational Procedures in *Cân Rolant*, the Middle Welsh Translation of *La chanson de Roland*', *Brathair*, 14 (2014), 109–28.

[134] Roberts, '*Ystoriaeu Brenhinedd*', p. 221.

[135] Russo, 'Translational Procedures', p. 123.

[136] *Selections from Ystorya Bown de Hamtwn*, ed. by Poppe and Reck, p. xx. See also Poppe and Reck, 'A French Romance in Wales 2', pp. 130–42.

[137] Poppe and Reck, 'A French Romance in Wales 2', pp. 142, 147–49.

[138] Poppe and Reck, 'A French Romance in Wales 2', p. 161.

ones did not occur in *Delw y Byd* because, not being a narrative text, it was not subject to the same degree of influence from local narrative conventions.

The translators of *Delw y Byd* were faced not only with unfamiliar place-names and personal names, but also with Latin constructions which have no Welsh equivalent. The Latin passive infinitives are a case in point. These have no Welsh equivalent and are often rendered in this text (Version A) as verbal nouns, e.g. *laureari* as *llaꝺenhau* and *contemplari* as *kytwledychu*.[139]

In terms of syntax, the Welsh appears to follow the Latin text very closely, even at the cost of diverging from normal Welsh grammatical conventions (particularly in Version A). However, as with the narrative texts, the translators of *Delw y Byd* do use native constructions, although they tend to do so where these closely parallel the structure of the Latin text. One example of a native construction used is from the introductory letter in the Red Book A Version: *Athro maꝺr y wybot a'e doethineb yn anuon annerch y athro arall...* The only difference from the Latin text is that the latter does not have a verb (a conventional feature of Latin epistolary style). Another example is the use of *yssyd y enꝺ* to render the Latin *est dicta*, 'is said', in RB-A 30[32]: *Yno Pontapolis y gan e pym dinas yssyd y enꝺ Beremoe, Asyone, Tolomaide, Apolinea, Cyrene, ac a gaꝺssant eu henꝺeu y gan eu hadeilwyr*, 'Thence Pontapolis, from the five cities is its name: Beremoe, Asyone, Tolomaide, Apolinea, Cyrene, and they got their names from their builders'.[140] Note that the Welsh translation reproduces the word order of the Latin original; the original Latin text reads: *Hęc et Pentapolis, a .v. civitatibus est dicta, scilicet, Berenice, Asrinoe, Ptolomaide, Apollonia, Cirene, a propriis conditoribus ita dictę*.[141]

Another example of a native construction, this time from Peniarth 17, is the phrase *Sef yu wybyr, llongeu y cawadeu*, 'This is clouds, ships of rains' (Pen.17 61[56]).[142] In this construction, common in Middle Welsh, *sef*, which always occurs in the beginning of a sentence, is the product of the copula *ys* ('it is') and third-person singular masculine pronoun *ef*.[143] The Latin is *Venti suo spiramine aquas in aera trahunt, quę conglobatę in nubes densantur. Dicuntur autem nubes, quasi nimborum naves*.[144] The Latin original relies on the sound

---

[139] Similarly, deponent infinitives also appear to be translated as verbal nouns, e.g. *imitari* as *gyffelybu*.

[140] Version B does not have the equivalent construction. White Book 30[32] reads: *Odyna y mae Sireneica gwlat a phym dinas yndi Brenice, Arsyone, Tholomoide, Apholinia, Cyrene.* The RB-B version is almost identical: *Odyna serenaica gꝺlat a phump dinas yndi: Brenice, Arsione, Tolomaide, Apollonia, Brenee.* The *ac a* in Version A probably represents a gapped reading; see discussion in note to Chapter 9[10], *Eil auon yꝺ Gyon, ac a elwir 'Nil'*, below, p. 82.

[141] *IM*, ed. by Flint, p. 63.

[142] See p. 62 below.

[143] *GMW*, p. 52.

[144] *IM*, ed. by Flint, p. 73; cf. RB-A 56: *Y gꝺynne hynny oc eu chꝺythyat a tynnant y dꝺyfyr yr*

similarity between *nubes* and *nimborum naves*, a similarity lost in the Welsh translation.

A recurrent construction which mirrors a specific Latin formulation is *yn y lle (y mae)*, used to introduce relative clauses.[145] According to Morris-Jones, it derives from *yn y lle y*, 'in the place that', and was often used to render the Latin *ubi*.[146] In our text its presence is almost exclusively dictated by the presence of a relative clause in the original (often as a variation on *in qua/hac est/fuit*). According to Morris-Jones, *yn y lle y* was the model for a similar construction, *yn yr hwnn y*.[147] Indeed, the use of *yr hwnn* etc. in translated works as proper relative pronouns is similar to the use of *yn y lle y* in *Delw y Byd*.[148] This may mean perhaps that in translated works a need for a relative pronoun/ construction was felt and one was artificially produced. Indeed, if *yr hwnn* is a later development, it would explain why, in contrast to the case of *y lle* in *Delw y Byd*, the relative use in cases examined by Russo for the narrative texts is not determined by the syntax of the original text.[149] This has implications for the relative dating of the translations, indicating that the narrative texts post-date *Delw y Byd*.

Finally, a construction that has already been the object of detailed study, as it is attested in many other Welsh translations from Latin, must be mentioned here: the use of *gwedy* + a verbal noun to represent the perfect passive participle.[150] There appear to be more instances in Version A, but Version A is also fuller, whilst Version B tends to abbreviate and paraphrase. Examples of this construction to be found in the text reproduced in the present edition include the two introductory letters (reproduced in the appendix), RB-A 31[32], Pen.17 65[60] and 97[91], and WB 65[60].[151] A fuller examination of all

---

awyr. ac yno y rewant ac y teơhaant ynwybyr. Yr wybyr a dywedir pan yơ llogeu kaơadeu ynt.

[145] There are sixteen instances of this construction in the RB-A version alone. It also occurs frequently in Peniarth 17 and the WB; S. Schumacher, 'Mittel- und Frühneukymrisch', in *Brythonic Celtic — Britannisches Keltisch: From Medieval British to Modern Breton*, ed. by Elmar Ternes (Bremen: Hempen, 2011), pp. 85–236 (pp. 208–09). This appears also to belong to the category described by Evans as pronominals used as antecedents; *GMW*, §74.

[146] Morris-Jones, *Welsh Syntax*, p. 101, §85 (ii) note.

[147] Morris-Jones, *Welsh Syntax*, p. 104, §86 (ii).

[148] *GMW*, p. 69 n. 1 and p. 66 n. 2, and Russo, 'Translational Procedures', p. 119. For more on use of *yr hwnn* etc., see, for instance, *Brut*, ed. by Roberts, p. liv; Luft 'Tracking ôl cyfieithu', pp. 171–72, 176.

[149] L. C. Russo, 'The Reception of Medieval French Narrative in Medieval Wales: The Case of *Chwedyl Iarlles y Ffynnawn* and *Cân Rolant*' (unpublished PhD Thesis, University of Buenos Aires, 2014), p. 141, 'Translational Procedures', p. 119.

[150] For more on this, see *GMW*, §217 (use b); Nicole Müller, *Agents in Early Welsh and Early Irish* (Oxford: Oxford University Press, 1999); T. J. Morgan, 'Braslun o Gystrawen y Berfenw', *Bulletin of the Board of Celtic Studies*, 9 (1938), 195–215; Ruth Carys Underdown, 'Studies in Welsh Prepositions: After' (unpublished PhD thesis, University of Cambridge, 2008).

[151] There are multiple other examples to be found in the text, including a number of more problematic cases, which require a more in-depth discussion, and are thus left to a separate study.

the translation techniques used in this text for rendering Latin perfect passive participles would perhaps yield some information on which techniques were adopted under which circumstances. One of the alternative constructions used is worth mentioning here because it is such a dominant trend in the text. This is the use of the Welsh preterite impersonal for the Latin perfect passive participle. It occurs in the multiple instances of *gelwit* for *appellata* or *ennwit* for perfect passive participle *vocata* < *voco, vocare*, 'call, name' (and in a few instances for *nuncupati* < *nuncupo, nuncupare*, 'call, name'). The use of the impersonal appears to be specific to these particular verbs, and its high frequency in this text is dependent on subject matter (it occurs in the descriptions of the origins of the names of countries, cities, and peoples). A more detailed study of other ways the text has of translating perfect passive participles is needed to establish what other contexts might be associated with precise techniques, but for now, the data on the use of *gwedy* + verbal noun shows that that construction in Welsh is the primary equivalent for Latin perfect passive participles. It is mainly used in cases other than references to origins of place-names (based on the verbs *appellare, vocare, nuncupare*), for which the impersonal construction is preferred. The features which distinguish sentences where *gwedy* + verbal noun is used from those where the impersonal is used remain to be determined based on a fuller study of the latter. The frequency of the use of other techniques for rendering the same Latin construction will need to be examined before a conclusion can be reached regarding the individual preferences of the translators.

There are also a number of lexical anomalies in the text which are explicable only as the results of the influence of Latin, such as *kyt-wledychu* (in the 'Letter of Christianus', printed at the end of the present edition), an attempt to translate, and probably influenced by the construction of, the Latin *con-templari*, the passive infinitive of *contemplo*, 'survey, behold, contemplate' (Lewis and Short, *Latin Dictionary*, s.v. *contemplo*).[152] The Welsh term appears to be composed of the prefix *kyt-*, 'co-' or 'joined' (GPC, s.v. *cyd-*), and what is probably *gwleddychu*, probably meaning 'to feast', perhaps in a spiritual sense, a meaning carried both by *gwledd*, 'feast', and *gwleddaf*, 'to partake in a feast, carouse, revel' (GPC, s.v. *gwleddaf*).[153]

To conclude, the text's syntactic and lexical features are consistent with

---

[152] *DB*, pp. 20–21. Another example of an unusual word form, possibly influenced by the Latin original, is *pressóyluodaf* in RB-A 20[21], glossed by Lewis and Diverres as 'inhabitatione'. The use is curious and appears to be constructed through the addition of a superlative ending to the noun *preswylfod*, 'dwelling place'. The superlative may indicate an attempt to render the Latin *inhabitatione gloriosa* in one word. The result appears to be a noun meaning 'dwelling' or 'residence'.

[153] For a discussion, see the note on p. 112 below.

what has been observed for medieval Welsh translated texts. The linguistic and orthographic features observed for the various manuscripts are similarly consistent with what we can expect on the basis of the manuscripts' provenance and with the date of the text. For instance, the later preposition 'to', *at*, is preferred to *ar*, which had been common earlier in similar constructions.[154]

Peniarth 17 has been the subject of previous studies and it remains to say here only that, as expected, its copy of *Delw y Byd* displays characteristics of early North Middle Welsh, including the use of *-th-* in the 3 singular pronoun version of the preposition *gan*: *ganthaб*.[155] The presence of 3 singular preterite ending in *-ws* (e.g. *rithus, symudws/symydws, emladws*) is also consistent with northern provenance and with the thirteenth-century date of the manuscript.[156] The most striking feature of the orthography of Peniarth 17 is the use of *e* for *y* (see below), which is also a northern feature. The northern features of the text in Peniarth 17 are indicative only of the northern provenance of the manuscript itself, and although a northern provenance of Version A would be consistent with what is known of the historical and cultural context of the text's transmission into Wales, the absence of these features in the RB-A version presents a problem.[157] The southern (probably Glamorgan) provenance of the Red Book is reflected in the text of the A version that it carries, for instance in its use of the southern form *whe*, 'six', with *wh-* instead of the standard *chw-* (RB-A 2[2]).[158]

It is unlikely that the A version was produced in the north and the language

---

[154] Cf. Mittendorf, 'Y Groglith', p. 269.

[155] In 135[129] (not in the present edition); cf. RB-A *gantaб* in 12[13].4, 12[13].16, RB-B 10[11].6, and *gantaw* in WB 20[21] (not in present edition). For more on the variation, see Peter Wynn Thomas, 'Middle Welsh Dialects: Problems and Perspectives', *BBCS*, 40 (1993), 17–50 (pp. 28–31), and '(-th-): Tystiolaeth Beirdd y Tywysogion a'r Uchelwyr', *Dwned*, 15 (2009), 11–32; Patrick Sims-Williams, 'Variation in Middle Welsh Conjugated Prepositions: Chronology, Register and Dialect', *Transactions of the Philological Society*, 111 (2013), 1–50 (pp. 32–42).

[156] Dialectally similar to Peniarth 14 (1–44); see Mittendorf, 'Y Groglith', p. 270. -ws is an early feature which was replaced in the North by *-awd* by the fourteenth century; see Thomas, 'Middle Welsh Dialects', p. 45, and Simon Rodway, *Dating Medieval Welsh Literature: Evidence from the Verbal System* (Aberystwyth: CMCS, 2013), pp. 128–53, at pp. 137–39, and ibid., p. 235.

[157] For more on the transmission, see Petrovskaia, '*Delw y Byd*' and 'Travels of a Quire'. Multiple examples of divergence between Peniarth 17 and Red Book A texts can be given, such as the retention of the diphthong *ae* in Peniarth 17, typical of thirteenth-century northern manuscripts (e.g. *caffael*, cf. *caffel* in RB-A, typical of fourteenth-century southern manuscripts); see Thomas Charles-Edwards and Paul Russell, 'The Hendregadredd Manuscript and the Orthography and Phonology of Welsh in the Early Fourteenth Century', *National Library of Wales Journal*, 28 (1993–1994), 419–62 (p. 419).

[158] *GMW*, p. 11. Another possible example is *yscaбnaf* in RB-A 3[3].8 and 3[3].11 and RB-B 3[3].5; cf. southern *ysgon* < *ysgawn* beside standard *ysgafn*, 'light'; *GMW*, p. 9, and *Culhwch ac Olwen: An Edition and Study of the Oldest Arthurian Tale*, ed. by Rachel Bromwich and D. Simon Evans (Cardiff: University of Wales Press, 1992), p. 155. For more on the Red Book and Glamorgan, see H. Fulton, 'The Geography of Welsh Literary Production in Late Medieval Glamorgan', *Journal of Medieval History*, 41 (2015), 325–40.

made consistent with a southern norm by the scribes of the Red Book or its exemplar, for there appear to be no indications of northern features in its language or orthography (if that were the case the odd slip to betray the ultimate provenance of the text would be expected). The question of whether the Red Book A text represents features introduced by its scribe (Hywel Fychan), or already present in its exemplar, is discussed below.

For Version B, whilst southern features dominate in the White Book and RB-B texts, the Philadelphia text carries some northern features. For instance, it has two instances of -th- with the preposition *gan* (and no instances with -t-): *ganthunt* (11[12]) and *ganthaϭ* (20[21]), and it also has *cibinnyeu* (< *cibyn*, 'shell'), with the yod in the plural.[159]

The scribal individuality that can be observed in the *Delw y Byd* texts is typical of medieval Welsh manuscripts, the orthographical features of which are the result of the practices of individuals rather than of a general system.[160] In the present edition, the orthography of the manuscripts is maintained to convey the full richness of the manuscript tradition. The glossary and indices at the back of the volume also follow the original spelling of the manuscripts, providing variants where the manuscripts spell the same word differently. The only alteration made for the purposes of the glossary is the removal of mutations.[161] The following brief overview provides a summary of the most significant cases of variation.[162]

Of the consonants, /k/ is variously represented by <c->, <k->, with some instances of <-ck->, <-cc->.[163] The grapheme <d> is used both for /d/ and for /ð/: e.g. *ansodedic* (for *ansoddedig*) in RB-A 33[34] and 74[69], or *arglϭyd* (for *arglwydd*) in RB-B 2[2]. Note that in terminal position <t> is also occasionally used for /d/: e.g. *byt*, 'world' (in all manuscripts), or *bryt* for *bryd*, 'mind' (in RB-A 'Letter of Honorius'). Generally, <th> is used to represent /θ/. The

---

[159] For more on yod as a northern marker, see Thomas, 'Medieval Welsh Dialects', pp. 26–28, 39. According to Thomas, the co-occurence of -th- and yod is a sign of a text's northern origins; ibid, pp. 39, 40.

[160] Paul Russell, 'Scribal (In)consistency in Thirteenth-Century South Wales: The Orthography of the Black Book of Carmarthen', *Studia Celtica*, 43 (2009), 135–74 (pp. 136–37).

[161] For more on mutations in medieval Welsh, see *GMW*, pp. 13–23; for a more general discussion of mutations in Welsh, see Martin J. Ball and Nicole Müller, *Mutation in Welsh* (London and New York: Routledge, 1992).

[162] For further detailed discussion of the relation of Middle Welsh orthography to the phonology of the language, see *GMW*, pp. xlv–14.

[163] There does not appear to be the same consistency in the variation of the use of <c> and <k> in initial position in these texts as that observed by Patricia Williams for the historical texts; *Historical Texts*, ed. by Williams (London: MHRA, 2012), p. xxxviii. Where a word occurs in lenited form in the text, the choice of initial letter for the non-lenited form in the glossary was based in each case on the predominant trend for that word in that manuscript. The results are perforce to some degree conjectural and any apparent trends should be treated with caution.

sound /f/ is consistently represented by <f> or <ff>, but <ph> is used in all manuscripts, in the following cases only: (1) to indicate spirant mutation of /p/, e.g. *phob* < *pob* (RB-A 1[1]), *phan* < *pan* (Pen.17 63[58]), *phriaϭt* < *priaϭt* (RB-B 74[69]); (2) place-names and other words taken from Latin, e.g. *Taphane* (RB-A 10[11]), *Pamphilia* (RB-A 20[21]), *Zephirus* (wind, in variant spellings, all manuscripts); and (3) the two animals *eliphant* (RB-A 12[13]) and *sarph*, 'snake' (WB 35[36]), pl. *sarphot* (WB 32[33]). The representation of /sk/ can vary within the same manuscript: <-sc-> or <-sg-> as in *yscriuennir* (RB-A 2[2]) or *ysgriuennir* (RB-A 52[47]).

Turning to the vowels, perhaps the most striking feature among all the manuscripts is the use of <e> for <y> in Peniarth 17, e.g. *Pyscaut ac adar a bressuyllyant en e duuyr* (56[51]). This is a thirteenth-century northern feature, where *e* was frequently used for /ə/,[164] as *y* was first adapted in South Wales and only subsequently made its way to the north.[165] In the manuscripts, the graphemes <i> and <y> (and sometimes <e>) are used interchangeably for /i/: *aniueil* / *anyueil* (RB-B 2[2]) and *aneueil*; *dineir*/*deneir* (both in WB 38[38]), while <y> is usually used to represent /j/: e.g. *caryat* (RB-A 10[11]). The greatest difficulty is presented by the graphemes <u>, <w>, <v>, and <ϭ>, which are used in these manuscripts for /v/, /u/, and /w/. The grapheme <u> is also used for /ʉ/ as in *ynteu* or *ugeint*. Similar variation is seen in the representation of /gʷ/ as <gw>, <gu>, <gϭ>, <gv>.

This brings us to the subject of the letter <ϭ>, a variant of <v>, and its use in editions.[166] There has been debate recently on the advisability of modernizing spelling with regard to this grapheme (as was done by Lewis and Diverres in *DB*). There is an argument that *ϭ* was perceived by medieval Welsh scribes as a separate letter, and therefore if one is to maintain the original orthography of the manuscripts, the use of this letter must also be kept.[167] Paul Russell has shown that Hywel Fychan uses <ϭ> as a separate letter, and therefore for both of the RB texts I use the letterform throughout the edition.[168] While there is less consistency in the uses of the letterform in the White Book and Peniarth 17 (the latter seems to use <ϭ>, <u>, <w>, and <v> interchangeably, for instance for /u/

---

[164] Russell, 'Scribal (In)consistency', p. 160.

[165] Peter R. Kitson, 'Old English Literacy and the Provenance of Welsh *y*', in *Yr Hen Iaith: Studies in Early Welsh*, ed. by Paul Russell (Aberystwyth: National Library of Wales, 2003), pp. 49–65; Mittendorf, '*Y Groglith*', p. 263.

[166] For the history of this letterform, see Charles-Edwards and Russell, 'The Hendregadredd Manuscript', p. 423.

[167] An extensive study of the spelling of /u/, /w/, and /v/ in Welsh manuscripts, including our texts, is to be found in Paul Russell, 'The Joy of Six: Spelling and Letter-Forms among Fourteenth-Century Welsh Scribes' (forthcoming). I am grateful to Prof. Russell for permitting me to consult a preliminary version of the study. See also Falileyev, '*Delw y Byd* Revisited', p. 74.

[168] Russell, 'Joy of Six'.

and /v/ in *glau, glaw, glaỽ; avon, aỽon; hwnnu, hvnnỽ, hỽnnỽ; hỽnn, hunn, hwn*), I have elected to keep the letterform in that text also. The reason for this is that in this case variation does not set it apart from the other letterforms (thereby there is no particular reason to remove only it).[169] If one were to normalize *hỽnnỽ* to *hvnnv*, for instance, it is unclear why *hwnnu* should be retained.

While the use of <ỽ> by Hywel Fychan, credited with setting 'an orthographical standard on the texts he was copying' which was followed by the other scribes of the RB, across both A and B texts he copied demonstrates consistency, there is also variation between the texts which suggests that some of the orthographical features pre-date the scribes of the manuscripts.[170] This point can be illustrated by comparing the use of <v> in Peniarth 17 to the use of <ỽ> by the same scribe in Peniarth 14 (1–44) and the Book of Aneirin.[171] The scribe of Peniarth 17 is the same as the so-called B scribe of the Book of Aneirin, and whilst his use of <ỽ> as a form of <v> is the cardinal feature distinguishing the A and B scribes of that manuscript, his use of <v> in Peniarth 17 is consistent with abundant evidence showing that he did not use the same orthographic conventions for both manuscripts. This reinforces the point made by Ingo Mittendorf that the poetry comprising the B-text of the Book of Aneirin is distinct from the Peniarth 14 (1–44) and Peniarth 17 prose texts not only linguistically but also orthographically, and therefore should not be used as a baseline for comparison.[172]

The variation observed in the orthographical practices of the *Delw y Byd* scribe of Peniarth 17 across manuscripts is also present in the work of Hywel Fychan. Preliminary examination has shown little evidence of normalization of the RB-A and the RB-B texts (both copied by him); they show almost the same amount of orthographic variation as can be seen between the WB and RB-B texts. For instance, a variant which can be observed in a comparison of the RB-B with the WB text, and which also emerges in comparison with RB-A, is the RB-B text's use of <i> where the other has <y>: *Ffison/Ffyson*, [*T*]*iger/tygris*, *Kaer Vadian/Kaer Madyan*, *Sirus/Syrus*, *Ismaelite/Ysmaelite*, *Sithia/Sythia*, *Ispania/Yspaen*. It should be noted, however, that these, and all other cases of this particular variation, are place-names. Simon Rodway has published an in-depth analysis of the orthographical differences between Hywel Fychan's

---

[169] Compare Hywel Fychan's consistency (e.g. *glaỽ, auon*). Another example of the difference in the use of the letterform between Hywel Fychan and the scribe of Peniarth 17 is their treatment of /au/. The Peniarth 17 scribe consistently gives *au*, while Hywel Fychan favours *aỽ* in the same words: e.g. *annyanaỽl, anyanaỽl/anyanaul; ansaỽd/ansaud*.

[170] Russell, 'Scribal (In)consistency', p. 136. It must be noted, however, that there is also a certain amount of inconsistency within the RB-A version of the text, e.g. in the spelling of /v/: *anniueil* and *aualeu* versus *anniveil*.

[171] See n. 95 above.

[172] Mittendorf, '*Y Groglith*', p. 263 n. 10.

version of *Culhwch* and the version of the same text preserved in the WB, which indicates that either Hywel or his exemplar was responsible for modernizing (to use Rodway's term) his text.[173] It may be worth undertaking a similar analysis for *Delw y Byd* to compare patterns of difference within the RB itself as well as between RB-B and WB texts. However, given the orthographic differences observed between the RB texts of *Delw y Byd*, it seems that at least for that text, orthographic intervention on the part of Hywel Fychan is unlikely.[174]

The additional fragment preserved in the Red Book, RB A-516, presents an interesting case. A comparison of this text with RB-A shows that the variation is minimal, indicating that they are closely related. An analysis of the orthographic features of the two texts for all five of the additional chapters (insofar as is permitted by the brief nature of the fragments) indicates that the spelling of the fragments tends to diverge from the affiliation of the text itself and correspond to the scribal preferences attested in the RB-B version, and in particular in doubling of <–n> to <–nn> and the occasional use of <k> for <c>. The orthographical continuity with the RB-B version suggests a common scribe at some stage prior to the current manuscript. It can also be concluded that the scribal features in all likelihood predate the RB scribe, since the RB-A and RB-B texts of *Delw y Byd*, which show considerable orthographic differences, were both copied by Hywel Fychan.[175]

The above overview has perforce been cursory, and further study of the orthography and language of the *Delw y Byd* fragments is needed. A comparison of the orthographic and linguistic features of the different versions may help throw light on the geographic origins and date of the versions (at least in the form in which they have come down to us), but is beyond the scope of the present work.

---

[173] Simon Rodway, 'The Red Book Text of "Culhwch ac Olwen": A Modernising Scribe at Work', *Studi Celtici*, 3 (2004), 93–161.

[174] For instance, there are instances of variation where *k* is consistently preferred to *c* by the B text, as in *Caṏcas/Kaṏcas*, *Caspium/Kaspiṏm*, *carṏ/karṏ*, *corn/korn*, *Cemageria/Kamegena*, *Canaan/Kynan*, for example. The only counterexample is *kufyt/cupyt*, 'cubit' with RB-A <k> to RB-B <c>. These instances are consistent with the trend observed in comparing the WB and RB. There is, however, at least one instance of WB <k> versus RB-A <c>: *kallavr/ callaṏr*. For more on the use of these graphemes see Simon Rodway, 'Cymraeg vs. Kymraeg: Dylanwad Ffrangeg ar Orgraff Cymraeg Canol?', *Studia Celtica*, 43 (2009), 123–33. Note that the exemplar of the B texts might not have shown lenition, since often only one MS marks it: e.g. *gyfarwyneb* (WB) / *kyfarwyneb* (RB-B), *y wlat* (WB) / *y gwlat* (RB-B); *chyn vryttet* (WB) / *chyn bryttet* (RB-B); *y vrydyon* (WB) / *yn brydyon* (RB-B); *keluydyt* (WB) / *gelvydyt* (RB-B). These examples are from chapters 29[31] and 32[33] of the present edition.

[175] Peter Wynn Thomas describes him as a 'low-noise, form-oriented scribe', who does not update or interfere in the text despite differences between it and his own dialect; 'Middle Welsh Dialects', pp. 21, 43.

Notes on the Commentary

Since Valerie Flint's edition already presents references to sources of much of the material in the *IM*, it seems superfluous to repeat that information here. The notes therefore aim to complement the information provided in the apparatus of the existing editions of the Latin and Welsh texts (Flint and *DB*).

The importance of *Delw y Byd* lies partly in the information it brought to medieval Wales and partly in the choices made by the translators, the knowledge of which can improve our understanding of medieval Welsh translation practices and understanding of the world, as well as what additional sources might have been available to the scribes or translators. Over and above that, we are presented with material that was available to and used by medieval Welsh authors. As Robert Wisnowsky observes, '[T]he material transformations that take place when texts are transmitted must be viewed as an important category of analysis, since they result not only from the intentions of translators or scribes but also from accidents of transmission.'[176] It is therefore necessary to understand which changes were intentional and which were not. Particularly interesting cases of intentional and unintentional changes are noted in the commentary to the text. These notes are by no means exhaustive, and it is hoped that the reader will find more useful information in the text and the variants presented by the manuscripts. All biblical quotations and references provided, unless otherwise noted, are from the Vulgate. In some cases I have provided the Latin text for comparison, and in all instances the English translations given, unless otherwise noted, are my own.

---

[176] Robert Wisnovsky et al., 'Introduction' to *Vehicles of Transmission, Translation, and Transformation in Medieval Textual Culture*, ed. by Wisnovsky et al. (Turnhout: Brepols, 2011), pp. 1–22 (p. 13).

CHAPTER LIST

The following presents the full contents of *Delw y Byd*, giving both chapter numbering systems. Angle brackets around chapter numbers indicate instances where the Welsh text does not yield itself easily to separation into chapters and is presented as a continuous paragraph in the present edition. Next to the chapter number, the abbreviations below are used to indicate the manuscripts in which those chapters are attested. Where only a small part of a chapter is attested in a manuscript, the manuscript reference is in brackets. The chapters included in the present edition are marked in bold, as are the manuscript sigla for those witnesses which are included in the edition for those chapters. The main manuscript used in the edition is always the first in the list. Chapters 4[4], 78[73], 120[114]–125[119], and 138[132] are not attested in any of the Welsh manuscripts.

RB-A — Red Book Version A
A-516 — the fragment of the A version preserved on f. 516 of the Red Book
RB-B — Red Book Version B
WB — White Book
Pen.17 — Peniarth 17
Ph. — Philadelphia 8680.O
Rawl. — Rawlinson 467
Add. RBA 248v — the bridging passage and additional fragment appended to the RB-A text

1[1] **RB-A, RB-B** (Intro., <1[1]>)
2[2] **RB-A, RB-B**
3[3] **RB-A, RB-B**
5[5] **RB-A, RB-B**
6[6] **RB-A, RB-B**
7[7] **RB-A, RB-B**
8[8] **RB-A, RB-B**
<8[9]> **RB-A, RB-B**
9[10] **RB-A, RB-B**
10[11] **RB-A, RB-B**
11[12] **RB-A, RB-B**, *Ph.*
12[13] **RB-A, RB-B**, *WB, Ph.*
13[14] **RB-A, RB-B, WB**
14[15] **RB-A, RB-B, WB**
15[16] **RB-A, RB-B, WB**
16[17] **RB-A, RB-B, WB**

18[19] *RB-A, RB-B, WB*
19[20] *RB-A, RB-B, WB, Ph.*
20[21] *RB-A, RB-B, WB, Ph.*
21[22] **RB-A, RB-B, WB, Ph.**
22[23] *RB-A, RB-B, WB, Ph.*
23[24] **RB-A, RB-B, WB, Ph.**
24[25] **RB-A, RB-B, WB**
<24[26]> **RB-A, RB-B, WB**
25[27] *RB-A, RB-B, WB*
26[28] *RB-A, RB-B, WB*
27[29] **RB-A, RB-B, WB**
28[30] **RB-A, RB-B, WB**
29[31] *RB-A, RB-B, WB*
30[32] *RB-A, RB-B, WB*
<31[32]> **RB-A, RB-B, WB**
32[33] **RB-A, RB-B, WB**

33[34] *RB-A, RB-B, WB*
34[35] *RB-A, RB-B, WB*
<35[36]> *RB-A, RB-B, WB*
36[37] *RB-A, RB-B, WB*
37[37] *RB-A, RB-B, WB*
38[38] *RB-A, RB-B, WB, Rawl.*
39[39] *RB-A, RB-B, WB, Rawl.*
40[40] *RB-A, RB-B, WB, Rawl.*
41[41] *RB-A, RB-B, WB, Rawl.*
42[42] *RB-A, RB-B, WB*
43[43] *RB-A, (RB-B), WB*
44[43] *RB-A*
45[43] *RB-A*
46[43] *RB-A*
47[44] *RB-A*
48[45] *RB-A, RB-B, WB*
49[45] *RB-A, RB-B, WB*
50[46] *RB-A*
51[46] *RB-A*
52[47] *RB-A, RB-B, WB*
53[48] *Pen.17*
54[49] *Pen.17*
55[50] *Pen.17*
56[51] *Pen.17*
57[52] *Pen.17, RB-A*
58[53] *Pen.17, RB-A, RB-B, WB*
59[54] *Pen.17, RB-A, RB-B, WB, (Rawl.)*
60[55] *Pen.17, RB-A, RB-B, WB, Rawl.*
<61[56]> *Pen.17, RB-A, RB-B, WB, Rawl.*
62[56–57] *Pen.17, RB-A, RB-B, WB, Rawl.*
63[58] *Pen.17, RB-A, RB-B, WB, Rawl.*
64[59] *Pen.17, RB-A, RB-B, WB*
65[60] *Pen.17, RB-A, RB-B, WB*
66[61] *Pen.17, RB-A, RB-B, WB*
67[62] *Pen.17, RB-A, A-516, RB-B, WB*
68[63] *Pen.17, RB-A, A-516, RB-B, WB*
69[64] *Pen.17, RB-A, A-516, RB-B, WB*
70[65] *Pen.17, RB-A, RB-B, WB*
71[66] *Pen.17*
72[67] *Pen.17, RB-A*
73[68] *Pen.17, RB-A*
74[69] *Pen.17, RB-A, A-516, RB-B, WB*
75[70] *Pen.17, RB-A, RB-B, WB*
76[71] *Pen.17, RB-A, RB-B, WB*
77[72] *Pen.17, RB-A, A-516, RB-B, WB*
79[74] *Pen.17, RB-A, RB-B, WB*
80[75] *Pen.17, RB-A, RB-B, WB*

81[76] *Pen.17, RB-A, RB-B, WB*
82[77] *Pen.17*
83[78] *Pen.17*
84[79] *Pen.17*
85[80] *Pen.17*
86[81] *Pen.17*
87[82] *Pen.17*
Add. RBA 248v Bridging Passage
88[83] *Pen.17, Add. RBA 248v*
89[84] *Pen.17*
90[84] *Pen.17*
91[85] *Pen.17*
92[86] *Pen.17*
93[87] *Pen.17*
94[88] *Pen.17*
95[89] *Pen.17*
96[90] *Pen.17*
97[91] *Pen.17*
98[92] *Pen.17*
99[93] *Pen.17*
100[94] *Pen.17*
101[95] *Pen.17*
102[96] *Pen.17*
103[97] *Pen.17*
104[98] *Pen.17*
105[99] *Pen.17*
106[100] *Pen.17*
107[101] *Pen.17*
108[102] *Pen.17*
109[103] *Pen.17*
110[104] *Pen.17*
111[105] *Pen.17*
112[106] *Pen.17*
113[107] *Pen.17*
114[108] *Pen.17*
115[109] *Pen.17*
116[110] *Pen.17*
117[111] *Pen.17*
118[112] *Pen.17*
119[113] *Pen.17*
126[120] *Pen.17*
127[121] *Pen.17*
128[122] *Pen.17*
129[123] *Pen.17*
130[124] *Pen.17*
131[125] *Pen.17*
132[126] *Pen.17*

TEXT

SECTION I. THE STRUCTURE OF THE WORLD

Red Book Version A

In RB-A, the first chapter is preceded, as in the original Latin text, by two introductory letters. The first is addressed to the author of the work, requesting that the encyclopedia be written. The response serves as an introduction to the text. The epistolary format, however, is limited to these first two letters, and in Version B (121v; col. 502), the introduction is limited to a short phrase introducing the title of the text.[1] Unlike the Elucidarium *and some other medieval works of similar nature, the rest of the* IM *does not follow a master-student dialogue structure. Rather, it is divided into chapters, each dealing with a particular phenomenon. The first describes the world as a whole.*

For the sake of comparison, text from RB-B is provided for the first three chapters.

1[1] Y byt a dywedir megys kyffroedic o bop parth, kanys yn dragywyda(l kyffro y mae. Y ffuruf yssyd ar lun pel, ac ar gyffelybr(yd wy yn dosparthedic (rth y defnydyeu. O diethyr ygkylch yr wy y mae plisgyn am y gwyn. Ac ygkylch y melyn y mae y g(ynn. A'r melyn yssyd ygkylch y defnyd bychan yssyd yn y perued. Val y mae byt y ffuruauent ygkylch yr a(yr glan. Odyna yr awyr glan ygkylch yr wybyr. Odyna yr wybyr ygkylch y dayar, ual y byd y melyn ygkylch y defnyd bychan.

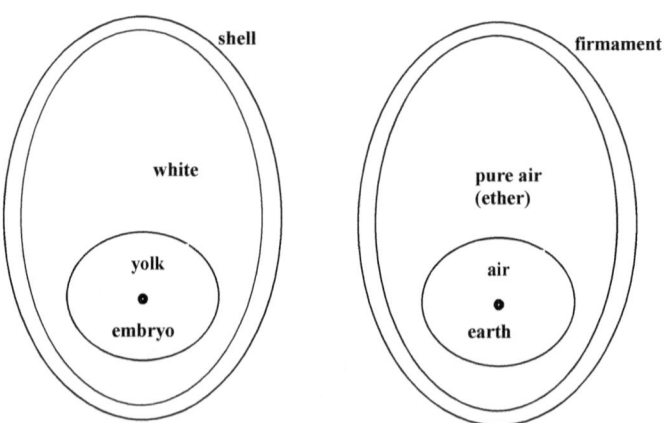

Figure 2: The egg metaphor and world structure according to Chapter 1

---

[1] The letters are reproduced below, pp. 74–75.

Red Book Version B: cols 502–16

The text in the B version begins with an introduction, added instead of the letters before Chapter 1.

Y llyuyr hỽnn a elwir 'Ymago Mỽndi'. Sef yỽ hynny 'Delỽ y Byt', kanys kedymdeith a'e gỽnaeth o arch y llall <1[1]> o ffuryfedigaeth y byt yr hỽnn yssyd ar weith pel gronn a gỽahannedic o'r defnydeu megys ỽy, megys y byd y kibynn ygkylch y Gỽynn, a'r gỽynn ygkylch y melyn, a'r melyn ygkylch y rith. Velly ~~megys~~ y mae y nef megys kibynn ygkylch y byt. A'r aỽyr pur yn y nef mal y gỽynn yn y kybynn, a'r aỽyr budur yn y gloyỽ mal y melyn yn y gỽynn. A'r dayar yn yr aỽyr budur ual y rith yn y melyn.

2[2] Creedigaeth y byt o bym mod yd yscriuennir. Vn onadunt: kynn amseroed ac oessoed yd oed holl ffuryf y byt y medỽl Duỽ. Ac y'r medỽl hỽnnỽ y dywedir: 'Y peth a wnaethpỽyt yndaỽ ef buched oed', yr eil greedigaeth pan grewyt y byt, ar gyffelybrỽyd medỽl Duỽ, megys y dywedir: 'y neb a pressỽyla yn tragywyd a wnaeth pob peth y gyt'. Tryded greedigaeth: pan grewyt pob ppeth yn wahanredaỽl herỽyd eu ryỽ yn y whe diwarnaỽt, ac o hynny y dywedir: 'Yn y whe diwarnaỽt y gỽnaeth Duỽ y holl weithredoed yn da iaỽn'. Pedwared creedigaeth uu: pan wahanỽys duỽ y creaduryeit pob un y ỽrth y gilyd, a phob un yn y greedigaeth yn amgen furyf, ual y gellit adnabot dyn a dyn, ac anniueil y ỽrth anniueil. Ac o'r greedigaeth honno y dywedir: 'Vyn Tat i a lauurya hyt yr aỽr honn, a minnheu a lauuryaf'. Pymhet greedigaeth yỽ: pan atnewydhaer ettwa y byt. Ac o hynny y dywedir: 'llyma y gỽnaf i bop peth o newyd'.

Red Book Version B

Pump creedigaeth a uu y'r byt. Kyntaf uu medỽl yr arglỽyd, a'r eil uu gyɲhebic y honno yn y defnyd, a'r trydyd pann ffurfỽyt y byt yn y dechreu oed, yn y whech diwarnaỽt. Pedweryd pan wahanỽyt dyn y gan arall, aneueil y gan aneueil, prenn y gan brenn. Pob vn o'e ryỽ ehun. Pymhet pan uynnỽyt enỽi y byt.

3[3] Ac yna y gỽnaethpỽyt y pedwar defnyd, a'r defnydyeu hynny yssyd ym pop peth. Nyt amgen: Tan, Awyr, Dỽfyr, Dayar. A'r rei hynny[2] a gerda pob un yn y gilyd yn eu kylch. Y tan yn yr awyr, A'r awyr yn y dỽfyr, A'r dỽvyr yn y dayar, a ymchoelir. Ac yg wrthỽyneb y dayar yn y dỽfyr, Y dỽfyr yn yr aỽyr, A'r awyr yn y tan a gedymdeithant. A phob rei onadunt oc eu priodolder a ymrỽymant pob vn a'e vreich dros y gilyd, ac a ymgymysgant yn gyfun bob eilwers. Kanys y dayar, sech ac oer yỽ. Ac a gytweda a'r

---

[2]   MS *hnny.*

dỽfyr oer. Y dỽfyr oer a gỽlyb yỽ, a gytweda y'r aỽyr gỽlyb. Yr awyr gỽlyb
a thỽym yỽ, a gytweda y'r tan gỽressaỽc a sych a gyuuna a'r dayar sech. A
chanys trymaf onadunt yỽ y dayar y mae yn issaf, a'r tan kanys yscaỽnaf
yỽ a achub y lle uchaf, a'r deu ereill,[3] nyt amgen y dỽfyr a'r awyr, yn y
kymherued y mae, megys rỽym kedernit. A chanys trymaf o'r deu yỽ y
dỽfyr, nessaf yỽ y'r dayar. A'r awyr kanys ysgaỽnaf yỽ nessaf yỽ y'r tan. Ac
yman y rifir yr aniueileit, a gerdont y dayar.

Red Book Version B

Ile y gelwit y defnyd y
gỽnaethpỽyt y byt ohonaỽ. Ac
y gwnaethpỽyt pedwar defnyd
corff. Ac y gỽnaetpỽyt pob peth
gỽedy hynny. Nyt amgen tan, ac
awyr, dỽfyr, a dayar, y rei a weda
pob un y gilyd. Kan yỽ sych y
dayar, ac oer; y[r] dỽfyr oer y gỽeda.
Dỽfyr oer yỽ a gỽlyb, y'r awyr
gỽlyb y gweda. Tan gỽressavc yỽ
a sych. Y'r dayar sych y gỽeda. O'r
rei hynn y dayar kanys trymaf
yssyd issaf. A'r tan kanys ysgaỽnaf
a gauas y lle uchaf, yr aỽyr yn
y perued, a'r dỽfyr nessaf y'r
dayar y llehawyt. Ar y dayar y
gỽnaethpỽyt petheu a ymdaant,
y'r dỽfyr y rei a nofyant, y'r awyr
y rei a ehedant. Y'r tan y rei a
lithrant.

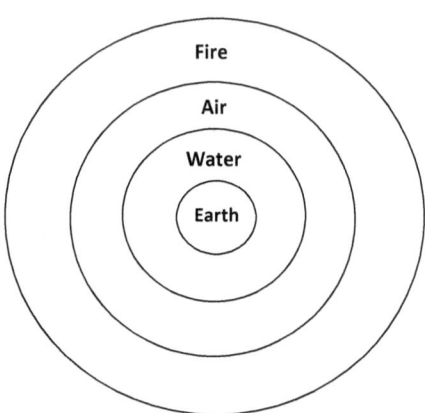

Figure 3: The four elements

*Chapter 4, on the seven names of
the earth, is absent from the Welsh
versions.*

5[5] Kylch[4] yr holl dayar a uessurỽyt, o riuedi un uilltir ar hugeint a mil o
vilioed o villtiroed deggveith, a phỽynt yssyd yn y kymherued y'r byt, mal
pỽynt yg kymherued kylch. Ac nyt oes dim yn y chynnal, yr holl dayar,
namyn gỽyrtheu Duỽ. Nyt a dros y theruyn mỽy noc vn o'r defnydyeu
ereill. Odyna y mae y mor ygkylch y dayar, mal amaerỽy, a fford trỽydi y'r
dyfred, megys gỽythi trỽy gorff dyn, y ardymheru y sychdỽr ym pob lle.

[3]   MS *erereill.*
[4]   MS *kyl.*

Ac ỽrth hynny pa du bynnac y clader y dayar, ef a geir dỽfyr yndi.

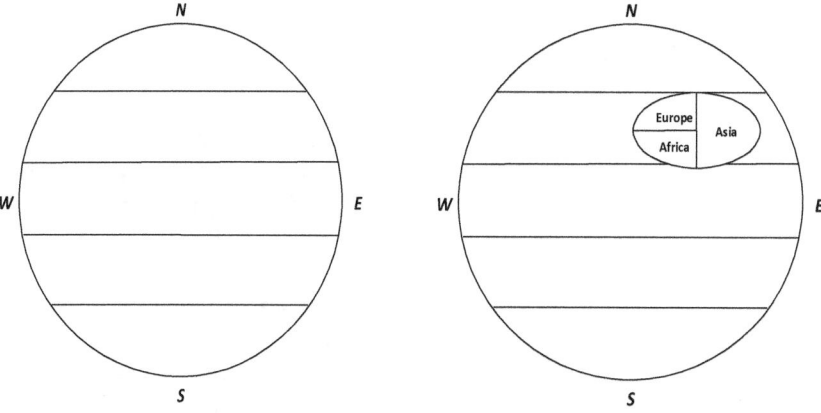

Fig. 4: The five zones (zonal map)

Fig. 5: Combination of T-O and zonal maps (approximation)

6[6] Yn bum rann y rennir yr holl daear, a'r dỽy rann eithaf onadunt, ny ellir eu pressỽylaỽ rac oeruel, a'r rann perued yssyd ampressỽyledic rac tragỽres. A'r dỽy gymherued yssyd ardymheredic o'r gỽres o'r neill parth, a'r oeruel o'r tu arall. Pei kneuit tan y gayaf pan uei oeruel a dan yr awyr noeth, ny ellir pressỽylaỽ yn y tan rac y wres; o bop parth y'r tan o bell, yn y lle nat ymgyʳhaedei wres y tan, y bydei y traoeruel. Ac y rỽng y tỽym a'r oer o bop parth y'r tan y bydei dỽy lin ardymheredic, o'r oeruel o'r neilltu. A'r gỽres o'r tu arall. Val hynny y gỽna yr heul. Kyntaf o'r ranneu hynny yỽ: septemtrionalis. Eil yỽ solsticialis. Tryded yỽ: equinoctialis. Pedwyred yỽ: brumalis. Pymhet yỽ: aỽstralis. Nyt oes yr ỽn bressỽyledic yni, namyn aỽstralis ehun.

7[7] A'r ran ardymeredic honno a rennir yn teir rann ygkylch Mor Groec. Vn yỽ yr Asia, arall yỽ Europa. Tryded yỽ yr Affrica. Teruyn yr Asia yỽ: o'r septemtrio trỽy y dỽyrein hyt a'r ueridiem. Europa yssyd o'r gorllewin hyt y gogled. Yr Affrica: o'r deheu, y lle a elwir 'meridies', hyt y gorllewin.

SECTION II: EARTH

The Tripartite World

Asia

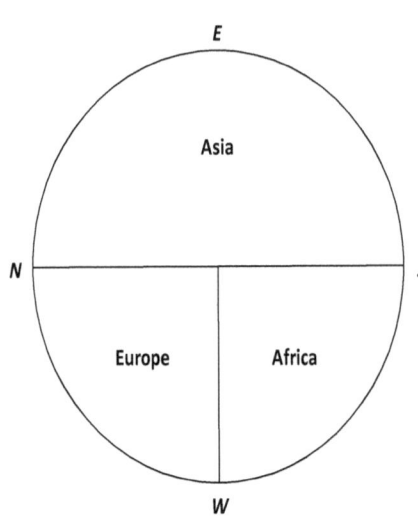

Fig. 6: The tripartite division of the
inhabited world (T-O map)

8[8] Yr Asia a gauaˢ y henỽ y gan Asia vrenhines. Ac o honno kyntaf brenhinyaeth yỽ yn y dỽyrein Paradỽys, lle kyflaỽn o bob kyfryỽ degỽch, a diffor^d y baỽp, kanys damgylchynedic yỽ o vur tan hyt y nef. <8[9]> Yno y mae prenn y uuched, a vỽytao ohonaỽ, yn yr ansaỽd honno y para byth. Yno y kyuyt ffynnaỽn o'r dayar. A honno a wehenir yn pedeir auon. A'r auonoed hynny a ant yn y dayar o vyỽn Paratwys, ac y deyrnassoed ereill ym pell yd ant.

9[10] Vn onadunt yỽ Ffyson. A honno heuyt a elwir Ganges. Ac a daỽ o Vynnyd Ocorbares yn yr India. Ac yn erbyn y dỽyrein y kerda yn y mor. Eil auon yỽ Gyon, ac a elwir 'Nil', ac a gyuyt o'r dayar gyr llaỽ Mynyd Athlans. Ac yn y lle yd a y'r dayar. Ac yn dirgel y kerda yn y dayar hyt yn traeth Mor Rud. Ac y kerda o newyd y ogylchynu gỽlat y Blammonyeit, a thrỽy yr Eifft y kerda ac y gỽahana yn seith aber y gyrchu ~~ffacta~~ y Mor Maỽr gyr llaỽ Alexand*r*ia. Tigris ac Euffrates y dỽy auon ereill, a gerdant o'r ffynnaỽn y tu ac Armenia, o Vynyd Caỽcas y llithrant, a thu a'r deheu y troant e eu hynt, ac y kyrchant y Mor Perued, yr hwnn a elwir Mor Groec. O'r tu hỽnt y Baradỽys y mae llawer difford amrauae^ul rac seirff ac anniueileit creulaỽn.

10[11] Odyna y mae gỽlat yr India, a gauas y henỽ y gan yr auon a elwir Indus. Ac a daỽ o septemtrio o Uynyd Caỽcas, ac y'r dehev y kerda, ac yd a ym Mor Rud. Teruyneu yr India yssyd hyt y gorlleỽin, ac o honno yd ennwir 'Mor yr India'. Ac yn y mor hỽnnỽ y mae ynys a elwir 'Taphane', arderchaỽc o dec dinas. Yn yr ynys honno y byd deu haf. A thyuu a wna yndi pob peth ym pob amser. Yn y mor hỽnnỽ y mae Crisa, ac Ergete, dỽy ynys ffrỽythlaỽn o eur ac aryant. Ac yn wastat y blodeuant. Ac yno y megir dynyon, ac y lliwir o'r lliw a dewissont. Yno y mae mynyded eur, ac rac

seirff ac adar y griffyt, ny ellir attunt. Yn yr India y mae Mynyd Caspius,
ac o h6nn6 yd enn6ir 'Mor Caspium', ac yr r6ng y mynyd h6nn6 a'r mor y
d6yrein y dywedir ry warchae o Alexander Ma6r kenedloed dywal gynt.
Sef oed y rei hynny: Gog, a Magog a 6wytteynt caraned dynyon, a chic
amr6t. Yn yr India y mae pedeir brenhinyaet a deugeint, a llawer heuyt o
boploed Coathras, Garmanos, ac eu hynyssed hyt yr awyr. Yn y mynyded
y maent Pigeneos, dynyon a deu gyfut yn eu hyt. Ac yn y dryded vl6ydyn
y magant, a'r wythuet yd henhaant. Ac eu nerth y6 ymlad ac adar y griffyt.
Ymplith y rei hynny y tyf y pybyr yn wynn. A phann loscer y mynyded y
ffo y seirff, y dua y gra6n hynny, ac y crycha gan y llosc. Yno y mae pobloed
a elwir Macrobyos deudec kyfut yn eu hyt a ymladant a'r griffonnyeit, yr
anniueileit y mae udunt corfforoed llewot, ac adaned, ac ewined, mal y
adar. Yno heuyt y maent y ry6 bobloed a elwir 'Agroctas' a Bragmanos, a
ant yn y tan y eu llosgi pob un o garyat y gilyd. Ereill yssyd yno a ladant
eu rieni g6edy hen^h aont, a g6neuthur g6led oc eu kic, ac ennwir y barnant
ar ny wnel uelly. Ac ereill yssyd yno a ymborthant ar bysca6t amr6t. Ac a
yfant heli y mor.

11[12] Yno y mae ry6 bopyl, a rann yndunt o dynyon, a rann arall o anniueileit.
Ac 6yth troet udunt. Ac eu g6adneu yn uchaf. Y mae ereill yno, a phenn
cr6nn arnunt. Ac ewined crymyon, ac yn wisc udunt cr6yn anniueileit.
A chyuarthyat k6n yn lle ^e u hymadra6d. Yno heuyt y mae y ry6 bobyl
a vagant bop bl6ydyn. Ac yn llwydon y genir. Ereill yssyd yno a uagant
yn y pymhet vl6ydyn. Ac ny pharhaant, o'r wythuet vl6ydyn allan. Yno
y maent ry6 bobyl vnllygeitya6c, ac a elwir 'Arismapi', a 'Siclopes'. Ereill
yssyd yno a seith troet udunt, ac o vntroet, buanach ynt, no'r awel wynt, a
thra orff6yssont ar y dayar, y dyrchauant yn wasca6t udunt, g6adyn vn oc
eu traet. Ereill yssyd yno heb penn udunt a elwir 'Lemennii'. Ac eu llygeit
yn eu d6yvron. Ac yn lle tr6yn a geneu udunt, deu d6ll yn eu d6yvron. A
g6rych arnunt mal ar anniueileit. Ereill yssyd geyr lla6 ffynna6n Ganges
auon. Ac nyt oes udunt ymborth, namyn arogleu aualeu. Ac yr pellet
y kerdont, ny byd udunt amgen noc aruein y fford y kerdont. A phan
gyfarffo dryc arogleu ac 6ynt y bydant veir6.

12[13] Yno y mae seirff, kymeint ac y llyngkant y keir6, ac y nofyant y mor.[5] Yno
y mae aniueil a elwir 'seucocreta'. Ac idav y mae corff assen, a d6y clun
car6, a d6yuronn a breicheu lle6. A thraet march,[6] a chorn ma6r holldedic.
A geneu llydan hyt y glusteu, ac un asc6rn yn garn^v an yn lle danned ida6,
ac ymadra6d dyn a uyd ganta6. Yno y mae aniueil arall, ac arthlenn march

---

[5] Punctuation mark in the MS '–'.
[6] Punctuation mark in the MS '–'.

idav, a d6yen baed coet, a llosc6rn eliphant, a chyrn kynnebonya6c idav.
A thra vo y neill tra'e geuyn, yd ymlad a'r llall. A phann bylo h6nn6, y try
y llall y ymlad. A du y6, a chystal y dicha6n ar vor ac ar tir. Yno y maent
teir6 melynyon, ac eu ble6 yn eu g6rth6yneb. A phenn dirua[6r] y ueint
udunt. A geneu llydan o'r clust y gilyd. Ac a ymladant ac eu kyrn bop
eilwers, ny mein[7] arnunt na chledyf na g6ae6, ac o'r damweina eu dala
ny ellir eu doui. Yno heuyt y mae anniueil a elwir 'manticora'. A gosged
dyn arna6. A their to ida6 o danned, corff le6, a llosc6rn Sarff. A llygeit
gleisson, a'e li6 yn goch. A ch6ibanat neidyr y symuda6 kywodolaetheu,
buanach yn redec, noc ehedyat yr ederyn. Kic dynyon vyd y ymborth. Yno
y maent ychen trichorna6c, a thraet meirch udunt. Yno y mae anniueil a
elwir 'monochero', a chorff march ida6, a phenn kar6. A thraet eliphant.
A llosc6rn h6ch. Ac vn corn ym perued y tal, o'e amdiffynn. A phedeir
troetued yn y hyt a dirua6r y6 y vlaenllymhet. Dywal y6 y b6ystuil h6nn, a
breuerat aruthur ganta6, ac a 6rth6ynepo ida6 a'e gorn y hergorn y herbyn.
Or delhir, y lad a ellir, ny ellir y doui. Yn Auon Gange y mae lassowot, a
thrychant troetued yn eu hyt. Yno y mae y ry6 bryuet kyffelyb y granc, [a]
deu vreich udunt o whech kufyt yn eu hyt. Ac a'r rei hynny y tynn atta6
yr eliphanyeit, ac y sa6d y weilgi. Ym Mor yr India y mae y ry6 genedyl o
valwot, ac oc eu kogyrneu y g6neir lluesteu didos y dynyon.

*The omitted chapters, 13[14] to 19[20], are (I use Valerie Flint's titles)
'Parthia', 'Mesopotamia', 'Syria', 'Palestine', 'Egypt' (which mentions Libya,
the Red Sea, the Nile, Thebes, and also locates Babylon in Egypt), 'Oriental
regions' (Caucasus and the region inhabited by the Amazons, Scythia,
Hunia, and Mt Ararat), and 'Asia Minor'. Chapter 15[16] locates Palestine
within the region it designates as Syria, and includes a discussion of
Jerusalem. Although often described as central to the medieval view of the
world,[8] it does not appear to be perceived as such in this text, though a
lengthy description of the origins of the name does attest to the importance
attached to it.*

---

[7] MS *men.*
[8] See, for instance, Iain MacLeod Higgins, 'Defining the Earth's Center in a Medieval
"Multi-Text": Jerusalem in the Book of John Mandeville', in *Text and Territory: Geographical
Imagination in the European Middle Ages*, ed. by S. Tomasch and S. Gilles (Philadelphia:
University of Pennsylvania Press, 1998), pp. 29–75; Simek, *Heaven and Earth*, pp. 73–81;
Philip Alexander, 'Jerusalem as the Omphalos of the World: On the History of a
Geographical Concept', in *Jerusalem: Its Sanctity and Centrality to Judaism, Christianity,
and Islam*, ed. by Lee. I. Levine (New York: Continuum, 1999), pp. 104–19. The context for
the non-central position of Jerusalem in this text is its non-central position in medieval
world maps; David Woodward has pointed out that the notion of the centrality of Jerusalem
in *mappae mundi* is based on a small number of famous examples; see 'Reality', pp. 515–17.

20[21] Bitinia a elwir Ffrigia Uỽyhaf, yn y lle y mae dinas Symina a adeilwys
Theseus. Odyna y mae Galathia a ennỽit y gan Gallis, y rei a elwis Bitinus
urenhin yn y ganhorthỽy. A gỽedy y uudugolyaeth honno yd edewis y
wlat. Nessaf y honno yỽ Ffrigya, a ennỽit y gan Ffrigius uab Europa, honno
a elỽit Ylyon y gan Ylon urenhin. A'r gaer honno a elwit Kaer Tro. Nessaf
y hynny y maent Licaonia, ac Ycaria, yn y lle y bu Auon Herme[s] a oed
glotuaỽr o diruaỽr eur. Odyna y mae Lidia, a ennỽit y gan Lidus urenhin.
Yno y mae Chyaỽra, odyna Saỽra. A'r gỽynt o bop parth udunt yn wastat.
Odyna y mae Cilicia a ennwit o enỽ y dinas a elwit Cilicia, a adeilwys Silix,
ac y gan hỽnnỽ y kaua[s] y deyrnas y henỽ. Yno y mae Mynyd Amana, ac
a elwir y Tarỽ. Yno y mae Dinas Tharsus a wnaeth Persus. Ac yno yd oed
pressỽyluodaf y Baỽl ebostol. Odyna Lithia, Passidia, Pamphilia. Odyna y
mae Ynys Bont brenhinyaeth llawer o genedloed, y'r lle hwnnỽ y diholet
Ovyd gynt. Ac odyna Clemens bap. Kann derỽ yr Asia, traeth6n weithon
o Europa,

Europe

21[22] a gauas y henỽ y gan Europ urenhin, neu y gan Europa verch Agenor.
Kyntaf teyrnas ohonei y tu a'r gogled y mae Mynyded Ris, ac Auon Tanais
a gauas y henw o Danay urenhin.

Red Book Version B

Evroppa gỽlat a gauas y henỽ y gan Europus vrenhin. Ac yn honno y
maent amryuael vynyded. Ac y mae Tanais auon a gafas y henỽ y gan
Tanaus vrenhin, ac yssyd o'r Mor Maỽr hyt yg kaer Teodosiỽm.

White Book

Europa gwlat a gauas y henw y gan Europ vrenin. Ac yn honno y maent
Ryphei mynyded. Ac y mae Thanais avon a gauas y henỽ y gan Thanais
vrenin. A Chorsicca yssyd o'r Mor Mawr hyt yn Theodosia.

Philadelphia

Europa gỽlad a gauas y enỽ y gan Europ vrenin, ac yn h[on]no y maent
y Mynyd Ripheri, ac y mae Thanais auon a gavas yr enỽ y gan a Thanais
vrenin, a Chosictu ysid o'r Mor Maỽr hyt yn Theodosia.

*The very brief Chapter 22[23], entitled in the Latin 'Of Scythia', explains that
Lower Scythia (given in Welsh as* Tithia Issaf) *stretches from the Don to the
Danube, and includes the provinces of Alania, Dacia (given as* Dascia), *and
Gothia.*

23[24] O Auon Danubi hyt yn Alpes y mae Germania Uchaf, a gauas y henỽ o
dyuyat y bobyl, kyfarwynep a'r gorllewin, Reno auon a'e theruyna, y tu a'r

gogled Auon Albia a'e teruyna. Yno y mae brenhinyaeth Sueuia, o vn enỽ a'e brenhin a elwit Sueuus. Yno y mae yr Almaen, a Recia. Yno y kyuyt Danubius auon o'r dayar, ac y honno y daỽ tri ugeint. Ac yn seith aber y gỽahanant y gyrchu Mor Ynys Pont. Yno y mae Narcus, a Banaza. Yn honno y mae Ratispona. Yno y mae Ffreinc dỽyreinyaỽl. Ac yn nessaf y honno Turnica. Odyna Saxonia.

Red Book Version B

O Danubius hyt ym Mynyd Mynneu y mae Germania, lle amlaf pobloed. Yn honno y mae Seusia brenhinyaeth, ac Almania. Ac yn honno y mae Danabius auon o'r dayar, a thrugein prif auon a a idi. Ac yn seith ab*er* yd aa yn i Mor Ynys Pont. Ac yn ho*n*no y mae Tiringa. Odyna Saxsonia.

White Book

O Danumbium avon hyt y Mynyd Mynev y mae Germania Vawr y lle amlaf y bobyl. Yn honno y mae Suesia brenhinaeth, ac Almonia. Ac yn honno y mae Danubius auon o'r dayar a thrugein prif auon a a yndi, ac yn seith aber yd a y Mor Ynys Pont. Ac yn honno y mae Ffreinc, ac y nessaf ydi y mae T*r*inga, a wedy hynny y mae Saxonia.

Philadelphia

[O] Danỽbiu[m] avon hyt y Mynyd Myneu [y … Ma]ỽr y lle amlaf y bobyl, yn hon[no] y mae Sues[…] breniniaeth ac Almania, ac [yn h]on[n] o y m[ae] Danvbius avon o'r dayar, a thrugein prif avon a a yndi ac yn seith aber yn [da … yn Mor] Ynyspont, ac yn honno y mae Freinc, ac y nessaf idi y mae Tringa, a gỽedy hyny y mae Saxonia.

24[25] O Albia y mae Germania Issaf, ar y tu a'r gogled, y mor a gae arnei. Yno y mae Denmarc, a Llychlyn ygkylch auon ᴰanubi, kyuarwyneb a'r dỽyrein hyt ym Mor Groec y mae Messia. Odyna Pannonia, a Uulgaria, <24[26]> odyna Tracia, o Chorstinobyl, a gauas y henỽ y gan Constantinus amheraỽdyr.

Chapter 25[27] (on Greece) and Chapter 26[28] (on Italy) are omitted here. The chapter on Greece enumerates the provinces, giving links between place-names and characters in the classical tradition (e.g. Aconia built by Helenus son of Hector) and includes references to Macedonia and Mount Olympus, amongst others. It misreads the Peloponnese as gỽlat Penelopensis and links the name to Penelope vrenhin (in Latin Peloponnese derived from the name of King Pelops). The chapter on Italy (described also as Groec Uaỽr) makes reference to Rome having been built by Romulus, and contains a description of important Italian cities which differs in order between the A and B versions of the Welsh text (discussed in greater detail in Petrovskaia, 'Delw y Byd').

27[29] Odyna y mae Gallia, honno a gerda o Vynyd Iubiter. Ac yn erbyn y gogled
y Uor Brytaen y teruyna. Honn a elwyt Ffreinc o enỽ Ffranckus vrenhin.
Pan doeth hỽnnỽ gyt ac Eneas o Tro yd adeilỽys ger llaỽ Renỽm, a Ffreinc
a dodes arnei. O honn y tu a'r gorlleỽin y mae Ffreinc Liỽn, a honno heuyt
a elwir Comaeᵗa. Odyna y rỽng Rodỽm a Liger y mae Gỽasgỽin.

Red Book Version B

Ac yno y mae Agallica, Bellica a daỽ o Vynyd Iouis parth a'r dỽyrein
hyt y Mor Brytaen. Ac yn honno y mae Ffreinc a enwit y gan Ffranckus
brenhin, gỽr a deuth o Tro y gyt ac Eneas Ysgỽydwyn.

White Book

Ac yno y mae Gallia Bellica. A honno a daỽ o Vynyd Iouis parth a'r
duyrein hyt y Mor Brytaen. Ac yn hon[no] y mae Ffreinc a enỽit y gan
Ffrancus vrenin, gỽr a deuth o Tro y genyt ac Eneas Ysgw[ydwl]yn.

28[30] odyna y mae yr Yspaen. Honno a gerda hyt y mor y gorllewin. Yn honno
y maent chwech brenhinyaeth: Terraconia, Cartago, Gallicia, Botica,
Tingỽirinia.

Red Book Version B

O honno parth a'r dỽyrein y mae Ispania, ac Iberia, ac Esperia. Ac yn
honno y mae seith cantref: Daraconia, Karthago, Lucitania, Gallicia,
Betica, Tignitania.

White Book

O honno parth a dỽyrein Yspania ac Yberia, ac Ysteria. Ac yn honno
y mae seith gantref: Terragonia, a Chartago, a Llucitammia, a Gallicia,
Betica, Tyngny[ta]nia.

29[31] Gyuerbyn a'r Yspaen parth a'r gorllewin yn y mor y mae yr ynyssed hynn:
Ynyˢ Prydein, Iwerdon. Gỽeryt y tir hỽnnỽ y ba wlat bynnac yd arweder a
lad y pryuet gỽennỽynic. Yn y lle y mae solsticỽm yr heul y mae Ynyssed
Orchades. Nyt amgen no dec ar hugeint. Yn Yscotlont y mae Tyle gwyd nyt
a byth eu deil y arnu. Yno y byd chwe mis haf a chwech gayaf. Ac chỽe mis
haf dyd yn wastat heb dim nos. A'r whech gaeaf, nos yn wastat heb dim
dyd. Y tu hỽnt y hynny parth a'r gorllewin y mae y mor rewedic, ac oeruel
tragywyd. Europam a gerdassam, aỽn weithon y'r Affric.

Red Book Version B

gyfarwyneb a'r Yspaen, parth a'r gorllewin ynyssed yssyd yn y mor:
Brytaen, Lloegyr, Iwerdon, Thanatos. A phrid y wlat honno a lad
anatred. Scothia Tilie, a gwyd y wlat honno ny byryant eu deil[9] vyth. Ac

---

[9] MS endeu.

yndi y byd chwe mis yn dyd yn wastat. A chwech ereill y nos, a thu hꝡnt y honno nyt oes dim namyn mor diffeith.

White Book

Kyfarwyneb a'r Yspaen parth a'r gvrllewin yn ynyssoed y mor y mae Britannia, Anglia, Hybernia, Thanatos. A phrid y gꝡlat honno **a** lad nadred. Socia, Tilie, **a** gꝡyd y wlat honno ny wywyant vyth eu deil. Ac yn honno y ꝡyd **.vj.** mis yn y dyd, **a** .vj. mis yn y nos yn wastat. Ac y tu hvnt y hynny nyt oes namyn mor diffeith.

Africa

30[32] Yr Affric a dech^r^eu yn dꝡyrein Indus auon. A thrꝡy y deheu a tynn hyt parth a'r gorllewin. Kyntaf gꝡlat ohonei yꝡ Libia, honn a gerda o Dinas Cerocemus. A Mynyd¹⁰ Cathalamus. Ac ym Mor Silen y teruyna. Ac o hꝡnnꝡ y dywedir 'Mor Libicꝡm'. Odyna Syrenaica a gauas y henꝡ y gan Syren urenhin. Yno Pontapolis y gan e pym dina^s^ yssyd y enꝡ: Beremoe, Asyone, Tolomaide, Apolinea, Cyrene, ac a gaꝡssant eu henꝡeu y gan eu hadeilwyr. Odyna y mae Tripolis o'r tri dinas y cauas y henꝡ. Nyt amgen: Occasa, Berete, a Leptis Vaꝡr. Odyna y mae [Bisace],¹¹ y gan y deu dinas y enꝡ. Odyna Adromeus, <31[32]> yn y lle y mae Cartago Vaꝡr a distrywys gꝡyr Ruuein. A gꝡedy y hatwneuthur y gelwit Cartago. Tewet y mur uu deu gyfut ar bymthe^c^. Odyna y mae Getulia, odyna Minidia. Odyna Dinas Yponensis, yn y lle y bu Aꝡstin yn escop, odyna Maꝡricia. Odyna Stipensis gꝡlat. Odyna Cesariensis.

32[33] Odyna yn y deheu y mae Ethiopia. Ac yn y dꝡyrein y mae Ethiopia arall, yn y lle y mae Kaer Sabba. Arall yssyd yn y gorllewin, yno y mae pobyl Troexdite yssyd uuanach no'r aniueileit. O'r tu hꝡnt y Ethiopia y mae lleoed¹² maꝡr diffeith, rac tragꝡres yr heul. Odyna y mae y mor, mꝡyhaf yssyd yn berwi yn wastat ual callaꝡr o wres yr heul. Yn yr ranneu eithaf o'r Affric, parth a'r gorllewin y mae dinas Gade^s^ a adeilvys y Phenites. Odyna y dywedir Mor Gadican, Athlas oed enꝡ gꝡreic brenhin yr Affric. Yno y mae Promethei y gan y rei y kymerth y mynyded y henꝡ. Yno y kaffat astrologia, o honno y dewir hyt ar y nef. Kann kerdassam ranneu yr Affric, mynagꝡn weithon yr ynyssed.

Red Book Version B

O hynny parth a hanner dyd y mae Ethiopia. Ac yn honno y mae Kaer Saba, a Mynyded Gargara. Ac yn yr rei hynny y mae ffynnaꝡn kyn

---

¹⁰ MS *Mynyde*.
¹¹ Place-name missing, and a blank space in the MS; *Bisace* supplied from Latin.
¹² MS *lleooed*.

oeret y dyd ac na eill neb y hyuet a chyn bryttet y nos ac na ellir mynet yn y chyỽyl. A chyfrỽng y mynyded hynny a'r dỽyrein y mae pobloed a elwir 'Trogatide'. A'r rei hynny a dalyant aniueileit gỽyllt o redec. Y tu hỽnt y Ethiopia y mae lleoed maỽr diffeith rac gỽres yr heul, a llawer o amryuael gen^edloed seirff nys atwen neb. Odyna y mae y Mor Maỽr a dywedir y vot yn brydyon mal kallaỽr o wres yr heul, yn eithaf yr Affric parth a'r gorllewin y mae Kagades a wnaeth Ffenices. Ac y genti y henwit Kadidamim. Ac yn y mor hỽnnỽ y mae Mynyd Athlans a enwit y ỽrth Athas vrenhin yr Affric. Ac yn y mynyd hỽnnỽ y gỽnaeth ef gelvydyt o'r ser. A'r mynyd hỽnnỽ a dywedir y vot yn kynnal yr aỽyr.

White Book

O hynny parth a hanner dyd y mae Ethi°pia ac yn honno y mae Caer Sabba. Yn y rei y mae Mynyded Gargara. Ac yn y rei hynny y mae ffynnawn kyn oere[t] y dyd ac na eill neb y hyuet. A chyn vryttyet y nos ac na eill neb mynet yn y chyvyl. Ac y rwg y mynyded hynny a'r dvyrein y mae pobloed a elỽir 'Trogadite', a rei hynny a d^aelyant anyueileit, gvyllt o redec. Y tu hvnt y Ethiopia y mae lleoed mavr diffeith rac gvres yr heul a llawer amraual genedyl sarphot, nys atwen neb. Odyna y mae Mor Mawr a hỽnnv a dyvedir y vrydyon mal kallavr o wres yr heul. Yn eithaf Affric parth a'r gorllevin y mae Ccaer Gades a wnnaeth Fenices, ac a enwit y genthi 'Cadidinum mor'. Ac yn y mor hvnnv y mae Mynyd Atlans, ac Atlas vrenin yr Affric a'e^13 gvnnaeth. Ac yn y mynyd hvnnv y gvnaeth ef keluydyt o'r ser, a'r mynyd hvnnv a dywedir y vot yn kynnal yr awyr.

Islands

This can be treated as a separate section that crosses the boundary between the three parts of the world described above, since all islands, regardless of location, are treated together in the following chapters.[14] The focus is on the Mediterranean.

33[34] Ynyssoed a dywedir oc eu bot yn ansodedic yn y mor. Y Mor Groec y mae Cyprys. Kyuerbyn a Sithia y mae Creta. Honno a ennỽir Centapolis. Honno yssyd gyuerbyn a Mor Libicum, Amdos yny^s, ac Hellesponto yn Europa. Edos ynys, ac a elwir Ciclades. Pedeir ynys ar dec ar hugeint yssyd wyneb yn wyneb. Vn yỽ Rodos y tu a'r dỽyrein. Odyna Tenedos ynys y tu a gogled y dỽyrein, odyna Carathos a'r dehev, hynny gyuerbyn a'r Eifft.

---

13 MS *a gvnnaeth.*

14 For a discussion of islands as a separate category in medieval geographical descriptions and further bibliography, see Nathalie Bouloux, 'Les îles dans les descriptions géographiques et les cartes du Moyen Âge', *Médiévales*, 47 (2004), pp. 47–62, and Silvère Menegaldo, 'Géographie et imaginaire insulaire au Moyen Âge, d'Isidore de Séville à Jean de Mandeville', *Les Lettres Romanes, 66 (2012), 37–86.*

Odyna ynys Tythera, y tu a'r gorllewin hỽnnỽ. Odyna Denos ynys ym perued Ciclade. Honno a elwir heuyt Ortigia. Odyna Ycaria ynys. Ac o honno y dywedir mor Ykareỽm. Odyna y mae Nawn, ynys Seint Daniel, odyna Ynys Varon, yno y byd y marmor gỽynaf, a maen sardin*us*. Odyna Sydon ynys, odyna Ynys Samos, odyno y hanoed Pitagoras, a Sibilla. Yno gyntaf y kaffat kywreinrỽyd y llestri dineu.

34[35] Odyna y mae Ynys Cilia, yno y mae[15] Mynyd Ethna, a brỽnstan yn llosgi yndaỽ yn wastat. Yn y moroed hynny y maent Karibdis a Ssill[e] perigleu moraỽl. Yno y buant[16] gynt y Ciclopes. Yno gyntaf y kaffat comedia. Geyr llaỽ Sicili y mae ~~mynyded~~ Ynyssed Eolye. Gyuarwyneb a Marsili y mae naỽ ynys a elỽir 'Stecades'. <35[36]> Ny enir yno na neidyr na bleid. Yno y mae anniueil a elwir 'solifuga' a lad dynyon ual ad[y]arcop, yno y mae llysseu apiascon, ac a lad dynyon dan chỽerthin, a chrydu eu gỽefleu. Yno y mae fynhonneu a uyd medeginyaeth y gleiuon, a delli y dynyon. Odyna y mae[17] Ynys Corsea, gyfuerbyn a'r Yspaen y mae Ynys Ebosus. Yno y ffoant y seirf. Yno y mae Ynyssed Baleares, a cheyr eu llaỽ Ynyssed Hesperide. Yno y mae amylder o deueit gỽynnyon, kystal eu gỽlan[18] a phorffor, y'r rei hynny y dywedir bot udunt knueu eureit. Y tu hỽnt y hynny y bu ynys uaỽr, ac a sodes a hi a'e phobyl pan yttoed Plato yndi yn yscriuennu. Ac a oed vỽy y medynt no'r Affric, ac Europa. A'r mor yssyd drosti yr aỽr honn. Meror yn ynys yn Auon Nil ym blaen Ethiopia, yno y mae prenn ebenus, a cher y lav y mae dinas Cyrene, yn y lle y mae y pydeỽ a wnaeth y doethon gynt o tri ugeint kyffut,[19] ac yn y waelaỽt y tywynna yr heul yg kymherued mis meheuin. Odyna y mae ynys yn y mor a elwir Perdita. Sef oed hynny colledic o bop ffrỽyth a thegỽch o neb ryỽ dim, pell y lles o neb ryỽ dayar, anednybygedic y baỽp or darffei o neb ryỽ damwein y chaffel, odyna pan geffit ny cheffit yno dim, yno y dywedit dyuot Brendanus. Yr ynyssed a gerdassam, traethỽn weithon o'r lle y mae uffern.

Red Book Version B

Silicia, Sicaria a Thrimacria, ac yn honno y mae Mynyd Ethna, a brỽnstan yndaỽ yn llosgi, yn y mor hwnnỽ y mae Silla, a Charibdis perigleu, <35[36]> a Sardinia ny byd yndi na bleid na sarff, namyn vn aniueil gỽyllt mal attyrcob, a'r neb a vratho a vyd marỽ. Ac yndi y mae llysseu tebic y palestyr, a'r neb a'e hysso a letta y enev mal dyn yn wherthin. Yn honno y mae ffynhonnev tỽumyn medeginyaeth y gleifon. Cursita, a Sirme, Sbesus ynyssed yn yr Yspaen.

---

[15] The third minim in *m* is a superscript correction.
[16] The initial *b* appears to be a corrected ỽ.
[17] MS *Odyna y mae y mae*. Repetition of *y mae*.
[18] MS *gỽlat*.
[19] *kuffut* in the MS. Single *f* in *DB* is probably normalised.

White Book

A Scisilia, a Trinaria. Ac yn honno y mae Mynyd Ethna, a brwnstan
yn llosci yndav. Ac yn y mor hvnnv y mae [Silla] a Charibdis periglev.
<35[36]> A Sardinia; ny vyd yndi na bleyd na sarph, namyn vn anyueil
[gvyllt] megys adyrcob. A'r neb a vratho hvnnv a vyd marv. Ac yndi
y mae llysseu tebic [y'r] [pa]llestyr a'r neb a'e hysso a leta y enev vegys
dyn yn chwerthin. Yn honno y maent ffynnhonev t'vym medeginyaeth
cleiuon. Corsica a Syrene. Ac Ebesus ynys yn yr Yspaen.

Hell

*Whilst Hell is commonly associated with fire, the present section still
concerns the element of Earth.*

36[37] MEgys y mae y dayar ym perued yr awyr. Val hynny y mae uffern ym
perued y dayar. Ac o hynny y dywedir hi y dayar eithaf. Lle aruthur yv o
dan, a brvnstan, ehang y waelavt, a chyving y eneu.

37[37] Honn a elwir heuyt 'llyn', neu 'dayar agheu', kanys marv uyd a disgynno
yno. Hon heuyt a elwir 'llynn tan', kanys ual y savd y maen yn y mor,
velly y savd yr eneideu yno. Honn heuyt a elwir 'dayar dywyll', kanys
tywyllvch mvc yssyd yndi, ac vybyr drevedic. Honn heuyt a elwir 'dayar
gyfyrgoll' neu 'dayar ebryuygedic'. Kanys mal yd ebryuygessant hvy duv.
Velly yd ebryuyka duv truga^rhau vrthunt vynteu. Tartarus heuyt, y gelwir,
o'e haruthder, a'e chrydvst, kanys yno y mae kvynuan a chrynua danned.
Hi ~~heuyt~~ heuyt a elwir 'dayar dan', kanys tyvyll yv y than ual dayar. Y
dyuynder a elwir 'erebus', llavn yv o'e myvn o dreigeu, a phryuet tanavl, a'e
geneu a elwyr baratrvm, kystal a morgervyn du. lleoed ereill yndi a elwir
'acheronta'. Nyt amgen, chwythyat yn taflu dryc ysprytoed. 'Stix' heuyt, y
gelwir, kystal a thristit. Auon yssyd yn uffern a elwir 'Fflegeton', aruthyr y
meint o wres tan. Ac anniodefedic o drewyant a mvc y brvnstan yn losci.
Yno y mae lleoed ereill o amryuaelon boeneu, yn y dayar, ac yn yr ynyssed
aruthur o oeruel,[20] ae ° wynnheu, ae o tan a brvnstan yn bervi yn wastat.
Kann derv in draethu o'r lleoed poenavl withyon, y traethvnn o'r dvfyR.

SECTION III: WATER

38[38] Y Dvfyr yv yr eil defnyd annyanavl, hvnnv a gynnullir yn uoroed. Ac
a wasgerir yn auonoed. Ac a rennir yn ffynnhonneu. Trvy auonoed y
rvymir. Trvy y dayaroed y gvesgerir. Trvy yr awyr y teneuheir, neu yd
hidlir. Yr holl dayar a rvym. Yr holl teyrnassoed a wahana, a phob rei

---

[20] MS *oruel.*

o'r kymydoed. Yr anodun a elwir yn eiga6n yn y lle ny chaffer beis. Nyt vnrym hagen y'r anodun ny bo gwaela6t ida6.

Red Book Version B

O'r d6fyr yssyd eil defnyd. Ac yn y mor y kynnullir, ac yn avonyd y dineuir. A thr6y y dayar y g6ehenir, a thr6y yr awyr y diedir, a'r holl dayar a r6ym. A'r holl wladoed a wahan.

White Book

A'r dufyr yssyd eil defnyd. Ac yn y mor y kynullir ac yn auonyd y dineir, a thrvy y daear y gvehenir, a thr6y yr avyr y deneir, a'r holl dayar a r6ym, a'r g6ladoed a wahan.

Rawlinson B 467

D6fyr yssid eil [d]efnyd, ac yn y mor y kynullir ac yn avenyd y dineuir a thr6y yr a6yr y dineuir a'r holl daear a r6ym yr holl vyd a [?]ana.[21]

39[39] Eigya6n kystal y6 ac amaervy g6regysseu. Pump gvre^{gi}s, neu bym rann y byt a damgylchyna y d6fyr, mal amwregis neu amaer6y.

40[40] G6regys y mor a uac lan6, a threi, pan y taflo y 6rthav y llein6. Pan y tynno atta6 y treiha. Peunyd y llein6 d6yweith, ac y treiha ual y tyuo[22] y lleuat, y tyf y llan6. Ac ual y kilio y lleuat y kilia y llan6. Pan el y lleuat y'r lle y bo kyhyt y dyd a'r nos, y byd g6ylltach y llan6 o nesset y lleuat. Pan el yr heul heuyt y'r p6nc uchaf y byd g6yllt heuyt y llan6 rac y phellet y 6rtha6. Tr6y un ul6ydyn eisseu o ugeint y teruyna y lleuat y chylch, ac o newyd y kymer dracheuyn.

Chapter 41[41] concerns chasms in the ocean, their role in the tides, and the existence of sub-terranean caves which capture winds.

42[42] O'r kyfry6 wynneu hynny y byd kyffro yn y dayar, kanys pan lauuryo y g6ynneu g6archaedic hynny y ymdianc odyno, y kyffroant y dayar o aruthred kyffro, ac y parant idi grynu.[23]

Chapters 43[43] to 47[44] discuss Sicily and its volcanoes, Mount Etna, Scylla, and the cold, which is said to derive from water as heat from fire.

48[45] Nyt achwanecka y mor yr a el ida6 o auonoed, kanys ual yd ymgymysco

---

[21] The last word is nearly illegible, but is probably *wahana*; the use of *vyd* attests to some divergence between the Rawlinson text and the other texts of the B version given here.

[22] The manuscript is repaired here and the words *y tyuo* are repeated.

[23] This is not followed by a capital letter in the manuscript, an indication that here our chapter divisions differ from the scribe's perception of the text.

y dỽfyr croeỽ a'r hallt y treulir, ae gan wythi y dayar, ae gan wres yr heul. Trỽy ffyrd dirgeledigyon yd ymchoel yr auonoed yndunt ehun.

49[45] Ac ỽrth hynny y para y mor yn hallt yr a el yndaỽ o'r awedỽr a'r glaỽ. Croewach uyd dỽfyr y ton no gỽaelaỽt yr eigaỽn.

50[46] Y Mor Coch a gerda o'r eigyaỽn. Ac eissoes lliỽ y dayar y kerda ohonei a'e gỽna ef yn goch; kanys coch yỽ y dayar. A'r tonneu hynny a liwha y dayar nessaf udunt.

Chapter 51[46] deals with the sea in general.

52[47] Eissoes deu ryỽ annyan yỽ: vn ʸ dỽfyr hallt a melys. Hallt uyd dỽfyr y mor. Melys ac yscaỽn vyd y dỽfyr a hannffo o bedeir auon paradỽys, mal yd ysgriuennir. Duỽ a duc pedeir auon, ac a'e gỽahanỽys y bedeir rann y byt, y ardymheru yr holl dayar. Ac y dywedir pan yỽ ohonunt yd henyỽ yssyd o ffynnaỽn ac auon o'r holl dayar. Ac y'r un ffynnaỽn ʸ ᵇᵃʳᵃᵈỽʸˢ yd ant dracheuyn, a chyt [?]²⁴yr awedỽr, na chymysc a dỽfyr y moroed, namyn llithraỽ o'r dỽfyr yscaỽn ar watʳthaf y trỽm, ac ymchoelut yna y eu fford dirgel. Ac ỽrth lithraỽ y dỽfyr croeỽ ar warthaf yr hallt, ỽrth hynny y mae melyssach y dỽfyr uchaf no'r weilgi yn y gỽaelaỽt.

RB-A omits chapters 53[48] to 56[51] (as does RB-B) and concludes the section on Water here with the words O'r dỽvyr y traethassam. weithon y traethỽn o'r awyr, *corresponding to the end of Chapter 57[52] (cf. the Peniarth 17 text below). Although after this omission the RB-A text continues on until Chapter 81[76], from this point onward in this edition the primary text is from Peniarth 17, since not only does it contain chapters 53[48] to 56[51], it also carries on continuously until the final chapters of the Latin Book 1. The text preserved in Peniarth 17 starts with Chapter 53[48], concerning hot water; 54[49] concerns waters poisoned by snakes.*

Peniarth 17

55[50] Paham na chyffry e [M]or Marw yr y guynnyeu ac na diodedᶠ endau dim byu, o achaus fynhonnyeu llynedic o byc o'r kyfryu byc ac yd adeilut Tur Babel ohonav, gurthuynep yu annyan y pyc y'r duuyr. Ac ny ellir torri y gueithret a wneler a'r pyc hỽnnỽ namen a guaet gureigyaul a gerdo o'e annyan ehun.

56[51] Pyscaut ac adar a bressuyllyant en e duuyr canys o henne y dywedir eu

---

²⁴ Letters illegible. *DB* suggests five letters in this gap; this is probable.

guneuth*ur*, er adar enteu a ehedant enteu en er awyr ac a bressuyllyant ar e daear o achaus bot er awyr en wlyp [m]egys y duuyr a'r daear enteu blith dra chemysc lau hep lau a'r duuyr. Paham enteu y gallant y ryu aniueillyeit pressuyllyau en e dyfred ac eu creu wynteu ar e daear ual y cocodrilli a'r moelronyeit? Urth uot ran endunt o'r duuyr ac o'r daear a bot en nessaf annyan y duuyr e'r daear.

57[52] Pan ymdywynyco y mor y nos y'r morduywyr mal tan temystyl a dau a phan weler y morỽoch yn ymgyuodi o'r tonneu henne y kyuyt guynt y'r awyr a'r wybyr kyffroedic ac a agorant yr awyr. E dyfred ry gerdassam, esgynnỽn weithyon e'r awyr.

SECTION IV: AIR

58[53] **O'r awyr.** Awyr yu pob peth o'r a welir en wac o'r daear hyt y lleuat o'r hun*n* y kynheliir buched paub o'r creaduryeit a chanys gulyp urth henne yd eheta endau adar ual y nouya y pyscaut en e duuyr. Endau y mae dieuyl ac eu poen arnadunt en arhos Dyd Braut ac o hỽnnỽ y kemerant corforoed pan ymdangossont y denyon.

59[54] Ohonav y creir y guynnyeu. E guynt yu awyr kynyruedic a chyffroedic ac nyt dim amgen a chuythat awyr, ac yn deudec y dosperthir ac enw priaut ar bop vn onadunt, a phetwar onadunt ysyd prifwynhyeu a'r lleill yn eskyllwynnyeu.

60[55] Kentaf o'r petwar prifwynt yu Septentrio, hỽnnỽ yu y gogledwynt hỽnnỽ a wna oeruel ac wyber, y tu deheu idau enteu Circius, a Thracias, a'r rei henne [a wnant] e eiry a chenllysc. E tu assw idaỽ Aquilo a Boreas, hỽnnỽ a wna rew en yr wybyr ac ya en e dyfred. Er eil prifwynt yu Subsolanus, gỽynt y gollewin yu hỽnnỽ, ac Affeliotes ardymeredic, y deheu enteu yu ỽulturnus a Chalcias a phob peth a sycha, y assw enteu wyber a ỽac. E trydyd prifwynt yu Auster, sef yu hỽnnỽ deheuwynt, hỽnnỽ a ỽac gulybur a mellt. E deheu enteu Euroauster, guressauc yu hỽnnỽ, y tu assw idau enteu Euronothus, ardymeredic. Guynnyeu y deheu a wnant y tymestleu muyhaf, en e mor canys o issel y chuythant. Y petweryd prifwynt yu Zephirus, Fauonius heuyt yu. Sef yu hỽnnỽ guynt duyrein a rydhaa y gaeaf ac a duc neu a arwed blodeu, y deheu yu guynt yr Affric, hỽnnỽ a ỽac tymestleu a tharaneu a mellt. O'r tu assw idaỽ enteu Chorus ac Argestes, wybyr a uac en e dwyrein ac eglurder en er India. Either henne y mae deu wynt ereill y ar y weilgi Aura ac Altanus, <61[56]> a'r rei henne oc eu chuythat a dynnant y duỽyr y'r awyr ac eno a dewheir yn gestyll ac wybyr. Sef yu wybyr llongeu y cawadeu.

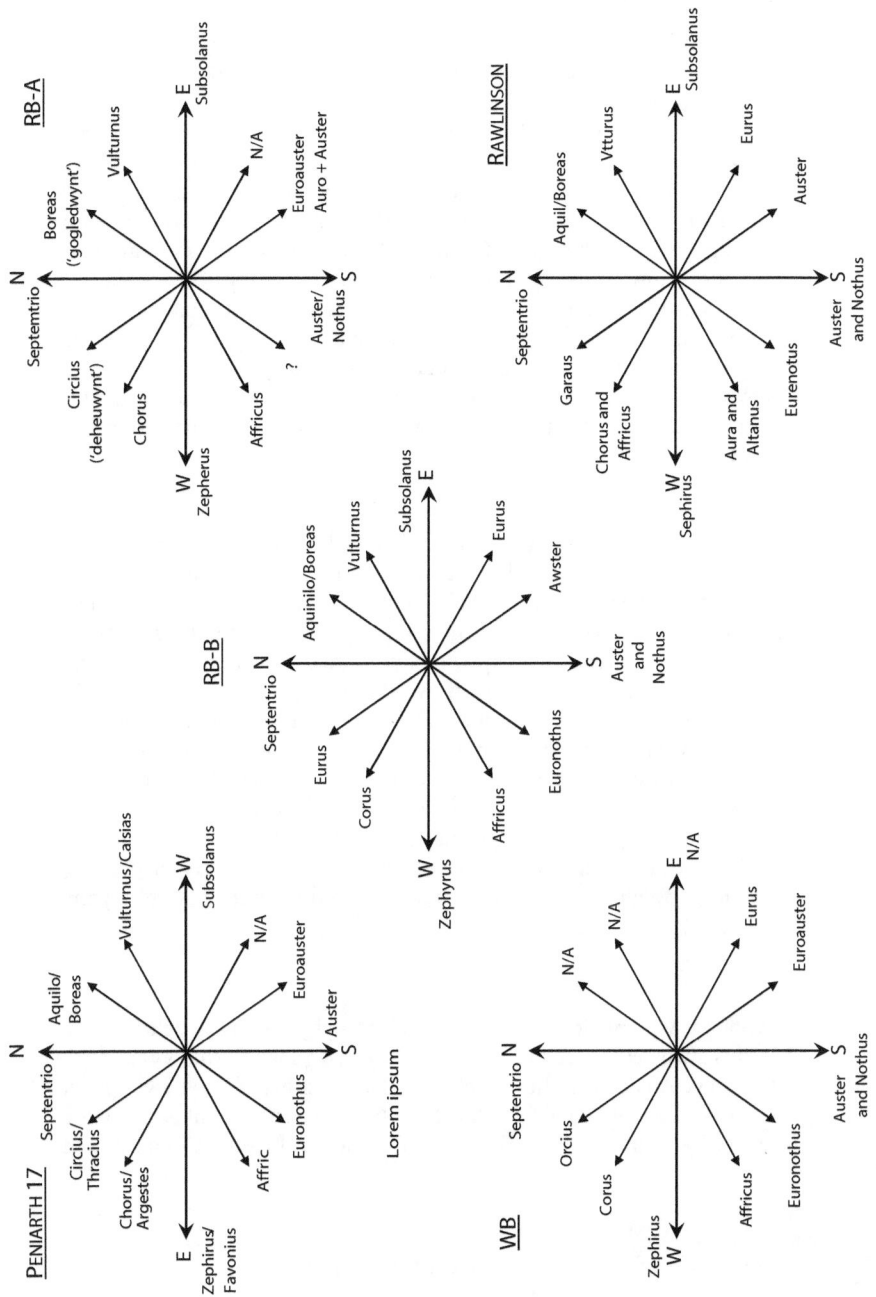

Fig. 7: The winds according to *Delw y Byd*

Red Book Version A

Kyntaf onadunt yỽ Septemtrio a wna oeruel ac wybrenneu.[25] Deheuwynt hỽnnỽ Circius, hỽnnỽ a wna eiry a chenllysc. Y gogledwynt yỽ Boreas, hỽnnỽ a geula yr wybyr. Eil prifwynt: Subsolanus, gwynt y dỽyrein, ardymheredic yỽ hỽnnỽ. Deheuwynt hỽnnỽ yỽ Uulteᵘrnus, pob peth a sycha, a'r gỽynt o'r tu asseu idaỽ ac a uac wybyr. Trydyd prifwynt yỽ Aỽster, ac a elwir Nothus, hỽnnỽ a uac gỽlybỽr a lluch, ac a wna llaỽer o tymhestleu. Ac ar y deheu y mae aᵉỽro. Aỽster gỽressaỽc yỽ, Euro ynteu gỽynt ardymherus yỽ. Gỽynneu Aỽster mỽyhaf[26] yỽ eu tymhestyl yn y mor, kanys o issel y chwythant. Pedweryd prifwynt yỽ: Zepherus, hỽnnỽ a uac gỽlith a blodeu, y deheu yỽ Affricus, hỽnnỽ a uac taraneu a thymhesteu, y asseu yỽ Chorus, yn y dỽyrein y mae wybyr,[27] ac eglurder yn yr India. Eithyr y rei hynny y mae deu wynt ar y weilgi a elwir Awra ac Altanus. <61[56]> Y gỽynne hynny oc eu chỽythyat a tynnant y dỽyfyr y'r awyr, ac yno y rewant ac y teỽhaant yn wybyr. Yr wybyr a dywedir pan yỽ llogeu kaỽadeu wynt.[28]

Red Book Version B

Kyntaf Septentrio, ac ef a wna oeruel ac wybyr. Ac o'r tu deheu y mae Eurus a wna eiry,[29] a chenllysc. Ac o'r tu asseu y mae Aquinilo, a Borias, y rei a dỽc yr wybyr. Yr eil yỽ Subsolanus, y gelwir ỽrth y gyuodi y lle y kyuyt yr heul, ac o'r tu deheu y mae Vulturnius a sycha pob peth. Ac o'r tu asseu y mae Eurus a wna nyỽl. Trydyd yỽ Aỽster a Nothus. Ac ef a wna llawer o dymhestloed yn yr awyr. Ac ef a wna gỽres, a gulybỽr, a mellt, a deheu yỽ, Awster gỽynt rỽʸmedic yỽ, ac o'r asseu y mae, Euromothus, a lleiaf a wna gỽynt y deᵐhestyl yn y mor. Kanys araf yỽ. Pedweryd yỽ Zephirus, a wna y blodeu a'r gỽlith, ac ef a dỽc y blodeu. Ac a yrr y gayaf, a dehev yỽ Affricus,[30] hwnnỽ a ỽna mellt ac asseu yỽ Corus,[31] hỽnnỽ a wna wybyr yn y dỽyrein, a goleuat yn yr India. A deheu ʸᵛ. Deu wynt ereill yssyd: Aỽra, ac Altanus. Aỽra, yn y dayar, ac Altanus yn y weilgi, <61[56]> a'r gỽynnoed hynny a yrrant y dyfred yn yr aỽyr, a'r rei hynny a rewhant yn yr wybyr.

White Book

Kyntaf yỽ Septemtrio ac ef a ỽna oeruel, ac wybyr. Ac o'r tu deheu y mae Orcius a wna eiry, a chenllysc. Ac o'r tu asseu y mae Eur[us a ỽna] nyỽl. Trydyd yỽ Auster a Nothus, ac ef a ỽna llaỽer tymhestyl yn yr awyr. Ac ef a ỽna gỽres a mellt a gvlybỽr. A dehev yỽ: Euro auster gỽynt tỽym yỽ. Ac o'r tu assev y mae: Euronothus. Lleiaf a ỽna gỽynt dehev o dymestyl yn

---

25   MS *wybenneu*.
26   MS *mỽyhaỽyhaf*.
27   Punctuation moved from after *dỽyrein* to after *wybyr*.
28   MS *ynt*.
29   MS *ery*.
30   Punctuation moved from before to after *Affricus*.
31   Punctuation moved from before *Corus* to after.

y mor kanys [a]raf yỽ. Petweryd [yỽ] Zephirus a ỽna blodev a gỽellt, ac a dwc blodev ac a yrr[32] gayaf. A dehev yỽ: Affricus a ỽna taranev a mellt. Ac assev [yỽ] Corus a ỽna [ỽy]byr yn y dỽyrein a goleuat yn yr Yndea. A deu wynt ereill Aura a Baltanus. Aura yn y daear, ac Vltanus yn y weilgi. <61[56]> A'r gỽynnoed a chrynnant dyfyred yn yr aỽyr. Y rei hynny a revant yn yr wybyr.

Rawlinson B 467

Kyntaf yỽ Septentrio ac ef ỽna oeruel ac ỽybyr oer. O'r tu deheu Arall a elỽir Garaus a ỽna eira a chenllysc. Ac o'r tu asseu y mae Aquil a Boreas y rei a dỽc yr ỽybyr. Arall yỽ Subsolanus, y gelỽir ỽrth y gyuodi yn lle kyffyt yr heul ac o'r tu deheu y mae ỽlturus a sycha[33] pob peth. Ac o'r tu asseu y mae Eurus a ỽna nyỽl. Trydid yỽ Aỽster a Nothus, ac ef a ỽna llaver o dymestloed yn yr aỽyr ac a ỽna gỽres a gỽlybvr a mellt. A deheu yỽ Aỽster gỽynt rỽymedic, ac o'r[34] tu asseu yỽ Eure Nothus. Lleihaa a ỽna gỽynt y deheu o dymystloed yn y mor kans araf yỽ. Pydỽryd yỽ Sephirus a ỽna blodeu a gỽlith ac ef a duc y blodeu ac a yr y gaeaf. Ac asseu yỽ Affricus a vna taraneu[35] a mellt, ac asseu yỽ Chorus a ỽna ỽybyr yn y dỽyrein a goleuat yn yr India. A deheu yỽ Deu ỽynt ereill: Aỽra ac Altanus, Aỽra yn y daear ac Altanus [yn y weilgi].[36] <61[56]> A'r gy[...] ed ar y[...][37] y [dyfr]ed yn yr aỽyr a re[i] [...] ynn[v] a a[38] yn yr ỽybyr.

Chapter 62[56–57] is on thunder and lightning.

63[58] E bwa en er awyr petwarlliwyauc o'r heul a'r wybyr pan del palader er heul a thywynnỽ yn wybren geu a phan urthuynepo yr wybren y'r palader ena yd ymdengys e bua hỽnnỽ, ỽal pei tywynnei heul ar lestreit duuyr tywynnu a wnaei en y urthuynep en e ty. O'r tan lliw tanaul o'r duuyr lliw coch, o'r awyr lliw guyrd, o'r daear lliw glas.

Chapter 64[59] deals with rain.

65[60] Y kenllysc enteu y defnynnyeu glaw ynt wedy rewi ac yaennỽ o oerỽel y guynnyeu en er awyr, ac eu symudaỽ en ỽein kenllysc.

Chapter 66[61] is on snow.

---

[32] MS *hyrr.*
[33] MS *sscha.*
[34] MS *ỽr.*
[35] MS *trayaneu.*
[36] The manuscript is illegible here and the text supplied. The beginning of the page, including Chapter 61[56], is largely illegible.
[37] Illegible. Probably five letters.
[38] Considerable gaps on both sides of *a*, but these appear to be blank.

67[62] E gulith a daw o'r awyr e'r daear; pan ỽo gorthrum o yrdder y nos
a lleuuer y lleuat y diskin ac y deiuyn. Os trechaf enteu uyd er oeruel
ymchuelut a wna y gulith en llwytrew neu en law.

Red Book Version A

Gỽlith a dygỽyd o'r awyr pan orthrymer o'r dyfred o echtywynnedigrỽyd
yr heul a'r lleuat y deiuyn. Ac os tyfu a ỽna oeruel y nos, ac ymchoelut y
gỽlith yn laỽ y gỽynha y gỽlith yn llỽytreỽ.

Red Book A-516 67[62][39]

Gỽlith a dygỽyd o'r aỽyr; pann ỽrthrymher yr awyr o'r dyfred o
echtywynnedigrỽyd y nos, ac ymchoelut y gỽlith yn laỽ, y gỽynha y gỽlith
yn llỽytreỽ.

Red Book Version B

A'r gỽynt a daỽ o'r wybyr.

White Book

A'r gỽlith a daỽ o'r aỽyr.

*Chapters 68[63] to 70[65] concern clouds, smoke, and meteors, termed
in Latin* igniculi, *'sparks'. It is worth noting that the text specifies that
the wind causes them to fall from ether into the air, posing a problem of
interpretation, for 'ether' was a complex concept which was interpreted by
various medieval writers in different ways. The Welsh translators chose to
render the term as* tan-defnyd, *'fire-element', identifying ether with fire as
the fourth element.[40]*

71[66] Tymestyl awyraul a ỽegir o drasychdur neu ormod gures er awyr ac a
gemero er awyr hwnnu ae en tynnỽ y anadyl ae enteu en buyta neu yn
yuet deissyuyt wenwyn a gymer a henne a ỽac y angeu. A dywedeis i oll y
a dan y lleuat y mae en er awyr y ar henne eglur yu hep dywylluc. Neu ry
gerdassam er awyr esgynnwn trostau enteu e tan.

SECTION V: FIRE

*The following section is dedicated to the element of fire. The A text is taken from
Peniarth 17. It should be noted that chapters 71[66] to 73[68] are lacking from the*

---

[39] Note that the A Version extracts from column 516 of the RB (following the B version) are
presented here in the conventional chapter order for the sake of convenience of comparison.
This order differs from the order in the manuscript, which runs: 74[69], 77[72], 67[62],
68[63], 69[64].

[40] See notes to Chapter 1[1] *O diethyr ygkylch yr wy* and *Odyna yr awyr glan ygkylch yr
wybyr*, Chapter 72[67] below, and note to that chapter under *pur awyr*.

B version.

72[67] E tan yr hỽn ysyd petweryd defnyd ny megir dim endau y derỽyn yu o'r
lleuat hyt y furuauent, a herwyd e mae gloewach a theneuach e duuyr no'r
daear, a theneuach a gloewach er awyr no'r duuyr euelly e mae teneuach
a gloewach e tan no'r awyr hvnnỽ a elwir y pur awyr ac a echtywynyca o
dragywydaul oleuat, o hỽnnỽ y kemer engylyon corforoed pan anỽoner ar
deneon.

73[68] Endaỽ e mae seith seren yg guahanredaul gylcheu en troi en erbyn e daear
ac am eu guibyauder redec y gelwir wynteu planedeu. E rei henne o diruaur
vuander y furỽauent a gribdeilir o'r duyrein hyt y gollewin, eissyoes oc eu
hanyanaul gyffro en erbyn y byt y prouir bot eu redec. Megys pei et ỽei
edyn en rot melin yr edyn eissyoes o'e briaut gyffro a laỽuryei en erbyn yr
olwyn y gerdet, y ỽelly y planedeu hyn en erbyn y sygneu.

74[69] Lleuat yu y kyntaf o'r planedeu a lleihaf o'r syr, a sef achaus hagen y guelir
hi en uwy no'r lleill urth y bot en nessaf e'r daear, en e kylch kentaf, corff y
lleuat ~~enteu~~ enteu ual pellen yu ac annyan tanaul a chymysc[41] duuyr ac ef,
ac nyt oes *priaut* leuuer ohonei ehun namen ỽal drych y ỽenfygyau e gan er
heul ac urth henne y gelwir hi lleuat ur[th] gaffael y lleuỽer y gan yr heul
sef yu henne goleuat. Er wybren a welir endi hitheu o annyan e duuyr y
credir. Ef a dywedir hagen pei na bei ~~gymy~~ gymysc y duuyr a hi e goleuhaei
y daear mal yr heul ac y bydei uwy y gure[s]saei no'r heul, y chorff hitheu
muy yu ac ehelaethach lawer no'r daear keny weler y bot hi en uwy no
guaelaut kerwyn am y huchet, y tu a ỽo y'r lleuat kyuerbyn a'r heul hỽnnỽ
a oleuhaa. A'r tu y urth yr heul hỽnnỽ a ỽyd tywyll, a phan ỽo pellaf y urth
er heul yna y byd kubyl oleu a chronn. Ny leihaa hi heuyt ac ny muyhaa
ac ny thyf ac ny chilya namen o gysgaut e daear y gu~~eduweir~~dwheir ac yd
anreithir o'e benfygyedic oleuat e gan er heul a chet kerdo hi beunyd o'r
duyrein hyt y gollewin o gymell y furuaỽent. Eissyoes yg gurthuynep y byt
y llaỽurya gan gerdet holl sygneu zodiacus yn wyth diwyrnaut ar ugeint. Y
chylch hitheu a gedernheir y adaỽ ohonei a'e gerdet en ỽn uluyden eissyeu
o ugeint. O byd coch y lleuat en betwared megys lliw eur guynnyeu a
darogan. Os en y chyrryeu y byd manneu duon dechreu y mis a dengys e
uot en glawauc. Os en y pherued y llaỽnlloneit yn eglur.

Red Book Version A

Kyntaf o'r seith blanet hynny yỽ y lleuat. Sef achavs y mae mỽy y gỽelet
no'r rei ereill, kanys nessaf y'r dayar yỽ y chylch, yn yr honn y mae

---

[41] MS *chymys.*

ansodedic yndi. Corff y lleuat pellenn y6 tana6l o anyan. A d6fyr yn y chymysc. Ac 6rth hynny nyt oes pria6t o leuat idi, namyn ual drych y lleuuerhaa yr heul[42] hi, a'r tywyll6ch a welir yndi ~~or d6fyr y goleu[?]y daear megys~~ ʸʳ heul o anyan y d6fyr y dywedir y bot. Ef a dywedir hagen pei na bei gymysc yndi o'r d6fyr, y goleuhaei y dayar megys yr heul, a cher 6yd y nesset m6y y g6ressaei. Kanys m6y o la6er a helaethach no'r dayar y6, kynn y bo m6y no g6aela6t baril y g6elet odyma rac y huchet. Y tu a vo y'r lleuat gyfar6yneb a'r heul a uyd goleu idi, a'r tu a vo y 6rthi a vyd tywyll. A pho pellaf y 6rth yr heul, yna y byd c6plaf y lleuat. Ny thyf heuyt nac ny leihaa, val y kilio yg g6asca6t y dayar o wyneb yr heul y kyll y goleuat. A chyt kerdo beunyd o'r d6yrein y'r gorllewin o gymhellyat y ffuruauen. Eissoes yn erbyn ~~yn~~ kyffro y dayar y llafurya y gerdet holl neu~~atus~~ yn wyth diwarna6t ar hugeint. Y chylch hitheu a dywedir y gwplau yn un ul6ydyn eisseu o ugeint. Os y lleuat y pedwyryd dyd o'r prif a goᶜha, ual lli6 eur, g6ynt a ardengys. Os yn y cornel uchaf idi y byd tywyll, ar6yd y6 y byd gla6a6c ar y th6f. Os yn y perued, arwyd sychin ar y lla6nlloneit.

Red Book A-516 74[69]

y tu a vo o'r lleuat gyvar6yneb a'r heul a vyd goleu idi, a'r tu a vo y 6rthi a vyd[43] tywyll. Ac atvo pellaf y 6rth yr heul yna y byd k6pla y goleuat. Os y lleuat y pedwyryd dyd o'r prif a gocha mal lli6 eur, gwynt a dengys. Os yn y korn uchaf idi y byd tywyll, ar6yd y vot yn la6a6c ar y t6f. Os yn y chana6l, ar6yd sychhin ar y lla6nlloer.

Red Book Version B

SEith planet yssyd, kyntaf y6 y lleuat, a lleihaf y6 o'r ser. Ac yn y kylch nessaf y mae. Ac annyan y tan a gymysc a'r d6fyr. Ac 6rth hynny nyt ytti6 y phria6t leuver genti. Namyn megyˢ drych y goleuhaa yr heul. A'r kysga6t a welhir yndi o annyan y d6fyr y6. Ac ef a dywedir pei na bei[44] y d6fyr, kyn oleuhet vydei a'r heul, a m6y y6 no'r dayar. Ac yn na6 niheu ar hugeint y kylcha y ffuruaven.

White Book

Seith blanet yssyd. Kyntaf y6 y lleuat a lleiaf y6 y syr. Ac vrth hynny y g6elir yn vvy no'r lleill rac y nesset y dayar. Ac yn y kylch kyntaf y mae. Ac o hannyan y tan yn kymyscu a'r d6fyr. Wrth hynny nyt ydi6 y phriaut leueuer namyn vegys drych y golevhaa. Y kyska6t a welir yndi o anyan y d6fyr y6. Ac ef a dy6edir pei na bei y d6fyr yndi kyn oleuet vydei a heul. A llet y6 no'r dayar. Ac y na6 mlyned ar hugein y kilya yr furuauen.

Chapters 75[70] to 85[80] are dedicated to descriptions of the individual planets (including the Sun, treated as one of the planets). The Saturn chapter, 81[76], concludes the RB-A, RB-B and WB versions of the text, and is reproduced below.

---

[42] MS *heu.*

[43] MS *a vyd a vyd.*

[44] MS *na bei na bei.*

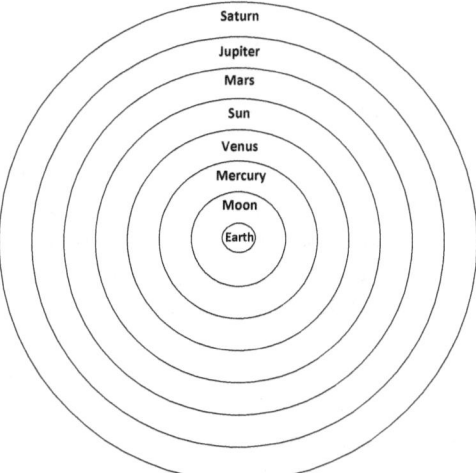

Saturn
Jupiter
Mars
Sun
Venus
Mercury
Moon
Earth

Fig. 8: The seven planets

81[76] Seithuet planet yu Sadurn, krwnn yw ac oeruelauc ac yg gurthuynep y byt y kerda y deudec sygyn trⱴy deng mlyned ar ugeint, a phuy bennac a dineuei delw o euyd yn dechreu Sadurn ef a brouit dywedut ohonei ⱴal den, ac wynt oll a guplaant eu kylch truy yspeit deudeng mlynedd ar ugeint a phymthecant.

Red Book Version A

Seithuet planet yⱴ Saturnus. Crⱴnn yⱴ, a reⱴlyt; yn erbyn y dayar, ac yn erbyn y plannedeu y danaⱴ a gerda trⱴy Zodiatus, trⱴy yspeit deudeng mlyned ar hugeint. Ac yn dechreu kylch Saturnus wedy deg mlyned ar hugeint. Pⱴy bynnac a dineuo delⱴ o euyd, hi a dyweit ual dyn. Paⱴb onadunt a gerdant eu kylch, wedy deudeg mlyned ar hugeint a phump cant. **A llyna diwed y llyuyr hvnn.**

Red Book Version B

Seithuet yⱴ Saturnus. A deg mlyned ar hugeint yⱴ y chylch. A pⱴy bynnac a wnelei delⱴ o evyd pan dechreuei y planet hⱴnnⱴ, gⱴedy deg mlyned ar hugeint, ef a allei brovi dywedut ohonei.

White Book

A phwy bynnac a welei delⱴ o efuyd[45] pan dechreue[i] hⱴnnⱴ gⱴedy dec mlyned ar hugeint, ef a allei dywedut proui ohonei.[46] A llyna diⱴed y llyuyr a elⱴir 'Ymago Mundi'.

---

[45] MS inserts *ef a allei* here, reading: *A phwy bynnac a welei delw o efuyd ef a allei pan dechreue[i] hⱴnnⱴ…*
[46] MS *ohonoei*.

The following chapters concern the apses of the planets, their colour, and the sign-bearing heavens (Lat. signifer). Chapter 84[79] describes the concept of the division of the heavens into the twelve parts of the Zodiac, and the rotation of the seven planets (including the Sun and Moon) underneath.

85[80] Y seith gylch henne a droant gan digriuaf gywydolyaeth, a'r gywydolyaeth honno ny daỽ atam ni canys uch er awyr e mae ac uch yu noc y dichaun an gallu ni y warandau ef. Ny allwn ni klybot dim o'r a ỽo y tu ỽry er awyr, o'r daear hyt y furuauent y messurir kywydolyaeth neuaul ac a'r dadogaeth honno y caffat an kywydolyaeth ninheu.

86[81] Gama en e daear. Odena ton en y lleuat, yr eil y Mercurius, y trydyd en ỽenus, y petweryd en yr Heul, y pymet y Mars. Chuechet en Iupiter. Seithuet yn Sadurn. Ac o henne y caffat messur y mussic. Seith ton a geffir o'r daear hyt y furuauen. O'r daear hyt y lleuat ton, odena hyt y Mercurius hanner ton. Odena hyt yg kylch y sygneu ton a hanner. En e ton y byd duy uilltir ar ugeint a chuechant a pymtheng mil o uilyoed o uilltiroed, en er hanner ton,[47] kemeint a hanner henne. Ac o henne y dechymygus e doethon naỽ awen canys o'r daear hyt e nef y caffat nau kyssondep a wedant y den en anyanaul.

87[82] Ỽal y mae dosparthyat y byt en seith ton ỽal henne ~~ỽal he~~ y dosperthir an keinyadaeth ninheu en seith lef, ỽegys henne y byd kyỽansodyat an ansaud ninheu, y corff o'r petwar defnyd a'r eneit o dri nerth ac o gyffelypruyd henne y kyỽansodet mussic o henne y gelwir den y byt lleihaf, ur^th ymgyffelybu o'e ansaud y neuaul organ.

88[83] O'r daear hyt y lleuat y mae ugeint milltir a chuechant a phymtheng mil o uilyoed o uilltiroed. O'r lleuat hyt y Mercurius deudeng milltir ac vyth cant a seith mil o uilyoed o uilltiroed a semis. Odena hit ar Uenus kymeint a henne. Odena hyt ar yr heul ỽn ỽilltir ar bymthec ar ugeint a phetwar cant a phedeir mil ar ugeint o uilyoed. O'r heul hyt ar ỽars, duy uilltir ar ugeint a chuechant a deudeng mil o uilyoed. Odena hyt ar Jupiter, deudeng milltir ac wyth cant a seith mil o uilyoed o uilltiroed. Odena hyt ar Sadurn y gymeint. Odena hyt y furuauen, ỽn ar bymthec ar ugeint a thrychant a phedeir mil ar ugeint o uilltiroed. O'r daear weitheon hyt y furuaỽen y mae pymtheng milltir a thri ugeint a thrychant a nau mil a chan mil o uilyoed o uilltired. E tan truy y planedeu ry esgynnassam, kerdun weitheon y petheu neuaul.

---

[47] Original punctuation mark ';'.

Additional Material from Red Book A, 248v; col. 998 line 27 to col. 999 line 17.
As discussed in greater detail in the introduction, the RB contains a further version of Chapter 88[83] which seems to belong to a text distinct from either the A version or the B version. It is preceded by a bridging passage that has no parallel in either the Latin text nor in any of the Welsh versions. The source of this transitional passage has so far eluded identification and it is possible that it was composed by a Welsh translator or redactor to serve as an introduction to Chapter 88[83].

[Bridging passage] **A** chyt traetho athraỽon gerd o'r sygneu yssyd yn Zodiacus, oduch y plannedeu. Ac o'r ffuruauen, ac o'r syr yssyd ossodedic yndi. Ac o'r nef y mae eistedua yr egylyon pob rei onadunt erbyn yn erbyn yn wahanredaỽl, ac eistedua duỽ drindaỽt y ar hynny oll. Nyt haỽd y neb corfforaỽl, na dyall peth kyfuch a hynny, na chyt as deallei athraỽon a lauuryynt yn hir yn dysc Llyfreu, ny ellynt hỽy eissoes eu menegi ỽy yn berffeith y leygyon, na gallu o leygyon y deall yr a ellit o uanac arnunt. Ac ỽrth hynny y tewit amdanunt.

88[83] **O'r** dayar hyt att y lloer[48] y maent pymtheg mil a whechant, a phymthec milltir ar hugeint. O'r lloer hyt y Mercuriỽm y maent: seith mil, ac ỽyth cant, a deudec milltir a hanner. O'r Mercuriỽm hyt att yr heul y maent: teir mil ar hugeint, a phetwar cant, ac un uilltir ar bymthec ar hugeint. O'r heul hyt att Uarten y maent: pymtheg mil a whechant, a phump milltir ar hugeint. O'r Varten hyt ar Jouen y maent: Seith mil ac ỽyth cant, a deugeint milltir a hanner. O'r Jouen hyt att Saturniỽm, y maent seith mil ac ỽyth cant, a deudec milltir a hanner. O'r Saturniỽm hyt att y ffuruauen y maent teir mil ar hugeint a phedwar cant. Ac vn uilltir ar bymthec ar hugeint. O'r dayar hyt att y nef y maent: can mil o uilioed. A naỽ mil o vilioed, a thrychant. A phymthec milltir ar hugeint.

Chapters 89[84] to 90[84] cover the various aspects of the heavens.

91[85] Pedeir rann ysyd y'r furuauen dwyrein a deheu, gollewin a gogled.

92[86] Duyrein a dywedir o achaus dwyrau yr heul ohonaỽ, e gollewin enteu a dywedir o achaus diguydau er heul eno. E deheu canys eno y byd hanner dyd, yn yaỽn, septemtrio enteu yu y gogled, a enwir o'r seith seren, a'r pedeir rann henne a elwir yg groec, 'Anatole', 'Dissis', 'Arcton', 'Messembria', ac o'r llythyren gentaf o'r petwar henỽ henne y kynullỽt enw Adaf sef yu hỽnnỽ y byt lleiaf.

93[87] Furuauen a enwir o'e bot en furyfaedic ym perued dyfred, a hỽnnỽ yu y

---

[48] Original punctuation mark ';'.

nef uchaf, cumpas o'e furyf dyfraul o annyan, ac yscythredic o syr en e amgylch oll. O'r duuyr e mae ual lliw yaen, neu 6al cristal.

94[88] Odena y mae duy werthyt y furuauen, 6n y neilltu y'r duyrein ac a welir y gennym ni, a'r llall en e deheu ny allwn ninheu byth y guelet, canys y daear ysyd y rom a hi, en y rei henne y try e nef 6al rot en y hachel.

95[89] Syr ysyd ossodedic en e nef en y holl gylch ac eissyoes nyt ymdangossant y dyd canys yr heul a or6yd ar eu goleuat, megys na thywynna yr heul arnam ninheu pan 6o yr wybyr en y chudyau. En rwymedic en e fur6auen y mae y syr ac ny diguydant namen kytredec ar furua6en o anryued 6uander.

96[90] Ac arn6nt yd edrech y morduywyr eu hynt ac eu kyuaruydyt, krwnn ynt wynteu a thanaul eu llunyaeth wynteu ac eu rinwed duw a'e guyr ehun er hwn [?] a wyr rif y syr ac eu enw, a'r sygneu a'r lleoed a'r amseroed. Doetheon y byt h6nn hagen ac eu henwis o enweu aniueillyeit a dynyon 6al y gallo denyon eu daall.

97[91] Yg kymerued e furuauen e mae deudec sygyn wedy ry lunyethu erbyn yn erbyn a doosparth truy gylch en gyuartal. Sef yu en6 y 'kylch zodiacus' y groec 'kylch y sygneu' yu henne yg kymraec.

98[92] Kentaf sygyn yu e Maharaen, llawer o syr ysyd endau, herwyd chuedyl h6nn6 yu y maharaen a'r croen eurin a duc Helles tros 6or ac a ducpuyt ymplith y syd[r], ac en lle sygyn y gossodet, canys megys y gorwed y Maharaen ar e tu deheu idau en hyt er haf a'r tu assw y gaeaf, y 6elly y byd er heul tra gerdo deheu y furuauen a dan y sygyn h6nn6.

There follow the descriptions of Taurus, Gemini, Cancer, Leo, Virgo, Libra, Scorpio, Sagittarius, Capricorn, and Aquarius.

109[103] Deudecet sygyn yu y Pyscaut, pan yt oed Antiocus urenhin en fo rac Uenus a Chupido e map yd emrithyassant wynteu en byscaut a guedy na beidyei e denyon e mor drwy hir amser rac eu llyngu o'r pyscaut y rithwt 6enus a Chupido en sygyn ymplith e syr, henne a arwydocaa bot er amser h6nn6 en glawauc.

Next come chapters dedicated to the various constellations: Pleiades, Arcton, Artophilax (Boötes), Arcturus, Phiton (serpens), Corona, Hercules, Lira, Cignus, and Aquila. The Welsh text omits chapters 120[114] to 126[120], which are dedicated to Cepheus and Casiopea, Perseus and Andromeda, Deltoton (= Triangulum), Serpentarius (= Ophiuchus), Pegasus, and Delphinus.

127[121] Odena e mae Sagitta a rodes Herculff y Philotetes trwy er honn y distrywyut Tro. E tu assw y zodiacus e tu a'r deheu e mae y sygneu hynn.

128[122] Idra caur a dec pen̲n̲ a deugeint idau a ladaud Herculff, ac o anryʋedaut y llehaut ymplith y syr.

There follow the descriptions of further constellations: Crater, Corvus, Orion, Canicula (= Canis Major), Lepus, Padus and Eridanus, Cerus, Centaurus, and Argo. Chapter 132[126], 'Anticanis' (= Canis Minor), which would have followed the chapter dedicated to Orion, is omitted in the Welsh text, as is Chapter 138[132], 'Ara' (which would have preceded 'Argo').

140[134] En eithaf e mae Pistrix neu Chimera, buystuil a phenn llew a chorff gauyr a lloscurn sarff, a ladaud Beloropho ac a ducpuyt ymplith y syr.

After a description of 'Canopus' described as seren yr Eifft, *'the star of Egypt', in 141[135], there follows an introduction to the concept of the 'hemisphere' in Chapter 142[135].*

143[136] Y guregys llaethaul a ʋyd guyn̲n̲ canys endau e dineu er holl syr en lleuuer.

Chapter 144[137] is dedicated to comets. Thereupon follows the description of the three heavens: the watery/aquatic heaven, the spiritual heaven, and, in the final chapter of the text, the heaven of heavens.

145[138] Y ar y fyrʋauen y mae dyfred ar gyffelybruyd wybren yn grogedic ac ygkylch y furuauen y credir eu bot en damgylchynedic, ac o hʋnnʋ y dywedir y nef dyfraul.

146[139] Y ar hʋnnʋ e mae y nef ysprydaul anetnebydedic y'r deneon. Eno e mae pressuylua guedy ry lunyethu yn nau rad. Eno e mae Paraduys y Paraduysseu en e lle y byd eneidyeu y seint. Hʋnnʋ yu y nef a yscriuenir y greu y gyt a'r daear en̲n̲ e dechreu.

147[140] Em pell odena y dywedir bot nef y neuoed en e lle y presuyllya Brenhyn er Engylyon.

APPENDIX I: INTRODUCTORY LETTERS

The following are the two introductory letters preserved in the beginning of the Version A text of the RB. The two letters are presented as a dialogue between student and teacher, a framework which Honorius employed in his Elucidarium *but which is not used in the* IM *beyond the two letters forming its preface.[1] For ease of reference, the names for the letters are given in English. Note, however, that these titles are not present in the manuscript.*

[Letter of Christianus] Athro ma6r y wybot a'e doethineb yn anuon annerch y athro arall kyfar6yd yn ffyd y drinda6t, ac yn ffrydyeu y doethineb, a'r keluydodeu o seith na6n yr Yspryt Glan, a g6edy gorffenno oessoed y uuched honn, lla6enhau o'r seith g6ynuydedigr6yd, ac yn yr wythuet kytwledychu, gyt a'r Tat Trinda6t yn vnda6t. Kann ytwyf yn ymberigla6 yn tywyll6ch ann6ybot, y gyt a'r rei ny wdant, 6rth hynny y g6elir y mi wybot mal dall yn d6yn vy muched yn drist gyueilornus. A chan atwen ynheu dy uot titheu yn damgylchynedic o diirua6r oleuat doethineb, yd 6yf y gyt ac ereill yn adol6yn g6rychonen o'th fflama6l wybot, pryt na leihaer hi itti, ac anuon ohonot attam ansodyat y byt, megys y my6n cla6r. Kanys truan yv pob peth o'r a 6el6n ni, wedy eu g6neuthur y rom, a ninheu[2] heb adnabot dim onadunt h6y, y gyt a'r anniueileit disyn6yr.

[Letter of Honorius] Y vab maeth y doethineb, yr h6nn yssyd yn ch6ilya6 yn garedic yg gordyuynder, ac yn anodun g6ybot grymhau o iechyt pob ry6 gyfry6 dyn, ac yn y nef, yn y lle y mae cudyedic tryzor y doethineb, edrych ar du6 wyd y gwyd, ac 6rth hynny pan vych di yn sychedic y sugna6 mer drud ~~yn yr yscrythur~~ yn yr yscruthur yn g6byl, ytt6yt titheu yn y deisyueit yn gedymdeithus, val y dyweit y diaereb: g6lanha ar auyr. Mi a ysgythreis itt ffuruf yr holl uyt, ual y gellych y welet a'th ol6c, ual y gallon y dyall, val y g6ahan6yt o'r defnydyeu kyntaf. A'r neges honno kyfla6n y6 o lauur a pherigyl, val yd atwaenost dy hun yn well no miui. Llauurys y6 ymi, yn achubedic yn llawer o betheu ereill, a molesteu uy mryt. Periglus y6 ynteu o acha6s dynyon kyghorvynnus ny orff6yssant yn aflonydu y petheu ny allont y gyffelybu, a'r petheu ny allont y gadarnhau, nyt ebryuygant y sathru mal buch dre6edic, a'r petheu a gablont ar gyhoed, a veuyryant dan gel. Ac oc an llauur ni a gymerant yn weithret udunt ehun, ac a sathrant ar gyhoed mal y mo[ch] y'r mererit. A chanyt vy llauur, namyn myhun oll, a dylyaf i y anuon

---

[1] For discussions, see de Ryke, 'Le miroir', p. 1252; Williams, *Fiery Shapes*, p. 80 n. 26.
[2] MS *niheu*.

ytti. Ac yn bennaf oll pryt na dyallỽyf i vyg geni y myhun, mỽy noc yr holl vyt gan ysgaelussaỽ y dynyon kyghoruynnus, ac a ymdodant ehunein, ac a gnoant ehunein o gallon gyghoruynnvs. Ac nyt miui a gnoant ỽy. A minheu a gyrchaf y goruchelyon defnydyon, kanys y llauur a orvyd ar pob peth. Ac ỽrth hynny ar dysc y lawer, nyt o amylder llyfreu udunt y gỽneir y llyfyr hỽnn. Ac y rodir yn enỽ idaỽ 'Delỽ y Byt' ⁰ achaỽs gỽelet yndaỽ llunyeith yr holl vyt y myỽn drych, yn yr hỽnn yd edᵉwir arỽystyl an kedymdeithas ni rac llaỽ. Ny dodaf i dim yndaỽ ef, onyt a ganmolont yr athraỽon mỽyhaf.

NOTES

1[1]

Y byt a dywedir megys kyffroedic o bop parth 'The world is called as if moving from all sides'. This translates the Latin *mundus dicitur quasi undique motus* (*IM*, ed. by Flint, p. 22), 'the world [= *mundus*] is called as though moved [= *motus*] from all sides'. Whilst in this case the etymological explanation present in the Latin text (*mundus < motus*) is lost, the use of *megys* in the translation usually signals a formulaic expression used by the translator to render etymological formulae that in the original Latin used *quasi*. For more, see Petrovskaia, 'La disparition du *quasi*'.

Y ffuruf yssyd ar lun pel 'Its [= the world's] form is like a ball's form'. This reflects the idea of the spherical earth; for more on medieval ideas of the shape of the earth, see Woodward, 'Reality', esp. pp. 518–19. Here, *yssyd* is 3 sg. ind. of *bot*, exceptionally used in a main clause in the so-called 'abnormal order'; see Schumacher, 'Mittel- und Frühneukymrisch', p. 216, and, for Modern Welsh, Peter Wynn Thomas, *Gramadeg y Gymraeg* (Cardiff: University of Wales Press, 1996), p. 501; for *yssyd* in *Delw y Byd*, see also the Introduction above, p. 34. There are a number of further instances of non-relative *yssyd* in this text. For instance, in *A'r melyn yssyd ygkylch y defnyd bychan yssyd yn y perued* in RB-A 1[1]4, the first *yssyd* is relative and the second non-relative. I am grateful to Erich Poppe for information on the use of *yssyd* in the abnormal order.

O diethyr ygkylch yr wy The author compares the concentric structure of the elements composing the world to that of an egg; see also Chapter 3. The egg simile usually contains only four elements. Here, seemingly a fifth element of ether is added (discussed in greater detail below), but the element of water appears to have been lost. This is already a feature of the Latin text and could be interpreted as an inadvertent result of Honorius's attempt to fit his five elements into the four-element metaphor of the egg. (See figures 2 and 3 above.) For more on the metaphor, see Simek, *Heaven and Earth*, pp. 19–22, and Chantal Connochie-Bourgne, '*Comment li element sont assis. L'image de l'œuf cosmique dans quelques encyclopédies en langue vulgaire du XIIIe siècle*', in *Les quatre éléments dans la culture médiévale*, ed. by Danielle Buschinger and A. Crépin (Göppingen: Kümmerle, 1983), pp. 37–48; for other examples in classical and medieval literature, see Peter Dronke,

Fabula: Explorations into the Uses of Myth in Medieval Platonism (Leiden and Cologne: E. J. Brill, 1974), Appendix A, pp. 154–66. The concentric structure of the elements described in this passage also reflects the structure of the book itself; see the discussion in the Introduction above, pp. 4–5. Note also the pattern of similes in this passage: A1, A2, A3, B1, B2, B3, A3 (where A represents the layers of the egg metaphor and B the corresponding layers of the world structure).

Val y mae byt y ffuruauent ygkylch yr aʋyr glan. This sentence is problematic and differs from the information in Version B. It is possible that *byt y ffuruauent*, 'the world of the firmament', refers to Heaven, since that is located outside the spheres; see *Delw y Byd* chapters 145[138] to 147[140] in the present edition.

Odyna yr awyr glan ygkylch yr wybyr. Wybyr would normally mean 'cloud' or 'sky', but here is used to translate the Latin *aer* (see *DB*, p. 117 n.7), probably to avoid using *awyr* which would lead to confusion with *awyr glan*, the term used by the translators for the Latin *purus ether*, 'pure ether'; cf. *IM*, ed. by Flint, p. 49. In an interesting departure from standard tradition, which identifies ether as the fifth element, in addition to Earth, Water, Air and Fire, Honorius, in *IM* Chapter 72[67], identifies Fire with ether: *hic etiam ether, quod purus aer dicitur, nominatur, et perpetuo splendore lętatur* (*IM*, ed. by Flint, p. 75), 'this is also named ether, which is called pure air, and rejoices in perpetual splendour'. He thus assigns to Fire the realm between the Moon and the firmament, which is conventionally associated with ether. For a discussion of the development of the idea of ether in the classical and medieval period, see Edward Grant, *Planets, Stars, and Orbs: The Medieval Cosmos, 1200–1687* (Cambridge: Cambridge University Press, 1994), pp. 14, 422–24.

1[1] Red Book Version B

Y llyuyr hʋnn This section has been included because Chapter 1[1] of this version begins halfway through its second sentence. It forms the simplified Version B equivalent of the complicated and lengthy introductory letters present in Version A (reproduced at the end of the present edition).

o ffuryfedigaeth y byt It is possible that this represents a title of the chapter, 'of the formation of the world'. The Latin chapter title (supplied by Flint, but also present in some manuscripts) is *de Creatione mundi*, 'On the Creation of the world'. In this case, if the phrase *Yr hwnn yssyd ar weith pel gronn* is to be taken as the beginning of a new sentence, it can be translated as 'This has the shape of a ball', with *yssyd* as 3 sg. pres. ind. < *bot*; for this form, see Schumacher, 'Mittel- und Frühneukymrisch', p. 216 §6. 4. 6, and *GMW*, p. 63 §66. This should be taken as continuing the introductory sentence of the text, rather than introducing a new chapter as in the modern convention for the text. As mentioned in the Introduction, in this and further such instances of continuity, the structure of the Welsh text has been maintained in continuing the paragraph, and the chapter number given in angle brackets within the text to facilitate comparison with other editions.

aϭyr budur This is a translation of the Latin *turbidus aer*; in Latin *turbidus* is commonly used to designate muddy, turbid, or thick liquids; see Charlton T. Lewis and Charles Short, *A Latin Dictionary* (Oxford: Clarendon Press, 1879), s.v. '*turbidus*'. For a discussion of types of air, see also note to *wybyr* above.

2[2]

y medϭl Duϭ This is an example of *yn* before *m-* becoming *ym*. The MS has *ymedϭl* for *ym medϭl*.

Ac y'r medϭl hϭnnϭ y dywedir introduces a biblical quotation (John 1. 3–4). The form *y dywedir* is used here and later in the text to translate a variety of Latin expressions which indicate quotations, citations, or references to general knowledge: *ut scribitur, sicut legitur, sicut scribitur,* and *sicut dicitur*; see *IM*, ed. by Flint, p. 49.

y neb a pressϭyla yn tragywyd a wnaeth pob peth y gyt Translation of the quotation from Eccl. 18. 1: *qui vivit in aeternum creavit omnia simul*, 'He who lives eternally created all simultaneously'.

Yn y whe diwarnaϭt y gϭnaeth Duϭ y holl weithredoed yn da iaϭn Marked as a quotation but not identified in *IM*, ed. by Flint, p. 49. Marked as a quotation in *DB* and identified as Gen. 1. This may be a paraphrase of Gen. 1. 31 or of Exod. 20. 9 or 20. 11. Mary Franklin-Brown considers the passage to be a conflation of the two quotations from Gen 1. 31 and Exod. 20. 9; see *Reading the World: Encyclopedic Writing in the Scholastic Age* (Chicago and London: University of Chicago Press, 2012), p. 102.

Vyn Tat i a lauurya hyt yr aϭr honn, a minnheu a lauuryaf In origin a quotation from John 5. 17: *Pater meus usque modo operatur, et ego operor*, 'My father works continuously, and I work'. The same quotation appears in the Welsh *Ystoria Lucidar*: *Vynn tat i a lauuryaᵬd hyt yr aᵬr honn. A mynnhev a lauuryaf* (RB, f. 34r, lines 4–5). In the *IM*, the relevance of this quotation to the main text is unclear, but it seems to illustrate the point that after the creation of man, he, in imitation of God, is also to work and create.

llyma y gᵬnaf i bop peth o newyd Quotation from Revelations 21. 5: *Ecce nova facio omnia*, 'Behold I make all things new'.

3[3]

pedwar defnyd The following is an articulation of the medieval doctrine of the four elements. The structure of the world described here provides the structure of the narrative of the text itself; see Introduction to the present edition, pp. 4–5 and note to 1[1], pp. 76–77.

Tan, Awyr, Dᵬfyr, Dayar Coloured initials in the names of the four elements correspond to the use of coloured initials to mark the corresponding sections in the text (in chapters 38[38], 58[53], 72[67]). Note that capitals are also associated with the elements in the following sentences, where the punctuation has been interpreted as commas rather than full stops to ease interpretation. For more on the structure of the text and the editorial decisions concerning punctuation, see the Introduction.

3[3] Red Book Version B

Ile y gelwit The first word is taken directly from the Latin without translation: *yle autem est materia* (*IM*, ed. by Flint, p. 50), '*yle* moreover is matter'. It is probable that *Ile* represents a chapter title, followed by a verb-initial sentence; a relative preverbal particle *a* would be expected here (with the phrase formulated, perhaps, *Ile a elwit*). Note that the two following sentences have a similar structure: verb-initial, introduced by *y*.

gᵬnaetpᵬyt Correctly *gᵬnaethpᵬyt*. The manuscript is missing an *h*; the correction over *p* identified by *RG* is probably a descender from the *p* above (and is in the wrong place).

Kan yᵬ sych y dayar, ac oer; yʳ dᵬfyr oer y gᵬeda Inserted *r* before *dᵬfyr*. This phrase is difficult. The punctuation should be read as a connecting rather than a dividing element, indicating that both dryness and coldness are properties of the earth. The phrase can be rendered as: 'because the earth is dry and cold, it is yoked/suited to the cold water' or 'because the earth is dry and cold, the cold water suits it/is yoked to it'.

5[5]

The measure of the Earth is given as 21,000,000, whereas in the original Latin it is 180,000 stadia, equated to 12,052 *milliaria* (*IM*, ed. by Flint, p. 51). The multiplicity of thousands might be due to the Latin use of *milliaria* for distance measures. The Roman *milliarium* is a mile of 5000 feet (where a foot was approx. 29.6 cm), thus equating to approximately 1480 meters; John Roche, *The Mathematics of Measurement: A Critical History* (London and New York: Springer, 1998), p. 24. The measure in the original text might be thus translated as approximately 17,837 km (real circumference *c.* 40,000 km). For the origins of the figure of 180,000 stadia for the Earth's circumference, see Irene Fischer, 'The Figure of the Earth — Changes in Concepts', *Geophysical Surveys*, 2 (1975), 3–54, esp. pp. 13–15. It appears that the stadium used in this text is one that corresponded to the Roman mile 15:1; ibid., p. 10.

6[6]

Yn bum rann y rennir yr holl daear The reference is to the zonal division of the inhabited world; see Woodward, 'Reality', p. 511. In a Welsh context, reference to the zonal division of the world is made in the Book of Taliesin poem *Kanu y byt mawr*, 'Greater Song of the World'; see *Legendary Poems from the Book of Taliesin*, ed. and trans. by Marged Haycock, 2nd edn (Aberystwyth: CMCS, 2015), p. 518; for discussion and references to further instances in medieval Welsh poetry, see ibid., p. 523.

Nyt oes yr vn bressoyledic yni, namyn aostralis ehun The Welsh version of the text contains an error: Honorius, as other medieval writers, considered only the zone they designated *solstitialis* inhabitable. The Austral circle, like the Septentrional, being close to the pole, was considered too cold for human habitation; for a brief and engaging discussion, see William Eisler, *The Furthest Shore: Images of Terra Australis from the Middle Ages to Captain Cook* (Cambridge: Cambridge University Press, 1995), esp. pp. 10–11.

7[7]

yn teir rann The tripartite structure of the inhabited world is reflected visually in the most dominant type of medieval world map, the T-O-type *mappae mundi*. An example from Wales is Exeter Cathedral Library MS 3514, f. 54r, where a T-O-type world map follows immediately on the *IM* text and is accompanied by an extract from Isidore on the twelve winds; see Crick, 'The Power and the Glory'. A similar and probably related map is present in Oxford, Jesus College MS 20, f. 32v, a Welsh-language manuscript which does not contain *Delw y Byd*. The division is also attested in medieval Welsh poetic texts. For instance, in *Saint a merthyron cred*, 'Creed of Saints and Martyrs', *Kanu y byt mawr*, 'Greater Song of the World', and Dafydd y Coed, 'Awdl i Dduw' ('Awdl to God'), lines 69–71; see *Blodeugerdd Barddas o Ganu Crefyddol Cynnar*, ed. by Marged Haycock (Abertawe: Cyhoeddiadau Barddas, 1994), p. 251; *Legendary Poems*, ed. and trans. by Haycock, p. 518; and *Gwaith Dafydd y Coed a Beirdd Eraill o Llyfr Coch Hergest*, ed. by R. Iestyn Faniel (Aberystwyth: University of Wales Centre for Advanced Welsh and Celtic Studies, 2002), pp. 35, 38.

Mor Groec = the Mediterranean; cf. *Bryt y Tywysogyon*, 'Chronicles of the Princes', entry for 1142, where Welsh pilgrims are referred to as drowned there; *Brut y Tywysogyon or The Chronicle of the Princes: Red Book of Hergest Version*, ed. and trans. by Thomas Jones (Cardiff: University of Wales Press, 1955), pp. 118–19; Petrovskaia, *Medieval Welsh Perceptions of the Orient*, p. 55.

8[8]

Ac o honno kyntaf brenhinyaeth yͣ yn y dͣyrein, Paradͣys Here *Paradͣys* is the subject of the sentence, with *brenhinyaeth* as predicative. There are punctuation marks before and after *Paradͣys*, and the position of the verb *yͣ* is changed in respect to the Latin: *Huius regio in oriente est Paradysus* (*IM*, ed. by Flint, p. 52), 'Of this the kingdom in the Orient is Paradise'.

9[10]

Ffyson Pison (one of the four rivers referred to in Genesis) and the Ganges appear to be regarded as one and the same river here. For more on the Ganges and the rivers of Paradise, see S. G. Darian, 'The Ganges and the Rivers of Eden', *Asiatische Studien / Etudes Asiatiques*, 31 (1977), 42–54, and J. T. A. G. M. van Ruiten, 'The Four Rivers of Eden in the *Apocalypse of Paul* 23. The Intertextual Relationship of Genesis 2.10–14 and the *Apocalypse of Paul* 23', in *The Visio Pauli and the Gnostic Apocalypse of Paul*, ed. by Jan N. Bremmer and István Czachesz (Leuven: Peeters, 2007), pp. 50–76, esp. 52–53. I am grateful to Elena Parina for bringing the latter to my attention. It is worth noting that these rivers are not present in the Welsh translation of the

Visio Pauli. In the *IM*, the identification of Pison with the Ganges, and the identification of Gihon with the Nile below, appear to represent an attempt to map biblical place-names and those transmitted from classical sources onto the same geographical system. The identification of the Gihon with the Nile is an old-standing tradition, attested in Jerome and Josephus among others; for references, see van Ruiten, 'The Four Rivers of Eden', p. 53 n. 12.

Vynnyd Ocorbares Latin original: *Orcobares.* These mountains are difficult to identify. C. Moseley, in his translation of Mandeville's *Travels* (Harmondsworth: Penguin, 2005), which also mentions these mountains, suggests tentatively that the reference might be to the Himalayas; p. 184.

Eil auon yϭ Gyon, ac a elwir 'Nil' Gyon and the Nile are identified as one and the same river. *Ac a* may be a form of relative, but rare (see *GMW*, p. 63 §65 n. 3). It is, however, more likely to be a gapped reading, with *Gyon* from the preceding content implied as the subject (null operator positioned between *ac* and *a*); see David W. E. Willis, *Syntactic Change in Welsh* (Oxford: Clarendon Press, 1998), pp. 124–26. There are other examples in this text of this phenomenon; see, for instance, 30[32] (*ac a gaϭssant*), 33[34] (*ac a elwir Ciclades*), <35[36]> (*ac a lad dynyon; ac a sodes a hi a'e phobyl*).

yn y lle Here used in the sense of 'immediately', 'at once', translating Latin *mox* 'immediately'. For a different use of *yn y lle*, translating *in qua*, 'in which', see note to 20[21] below (*yn y lle y mae*).

gϭlat y Blammonyeit See Petrovskaia, *Medieval Welsh Perceptions of the Orient*, pp. 12–13. This translates the Latin *Ethiopia*; *IM*, ed. by Flint, p. 52.

ffacta Deletion probably by the original scribe. It is unclear how this word came into the manuscript. It may represent the Latin text from which the Welsh was translated. However, the word *facta* does not occur in the Latin in this section of the text.

Mor Maϭr This appears to be a reference to the Mediterranean. The name *Mare Magnum*, for which *Mor Maϭr* is an exact translation, is also used sometimes on T-O world maps as a label for the Mediterranean.

10[11]

gϭlat yr India For the association of India with marvels, see, for instance, Chet Van Duzer, '*Hic sunt dracones*: The Geography and Cartography of Monsters', in *The Ashgate Research Companion to Monsters and the*

Monstrous, ed. by Asa Simon Mittman and Peter J. Dendle (Farnham: Ashgate, 2012), pp. 387–435.

ac y lliwir o'r lliw a dewissont This reference appears to be absent in most manuscripts of the *IM*; it is added in the Corpus MS: *Quę insulę ginnunt homines tinctis cordibus* (*IM*, ed. by Flint, p. 53 n. 10.6), 'Which islands bear men with coloured hearts'. This variant was not known to Lewis and Diverres, see *DB*, p. 119 n. They identify the source of the reference in Isidore's *Etymologiae*, XIV. 3. 6. However, Isidore refers to *tincti coloris homines*, 'men coloured of appearance', and *cordibus* in the Corpus manuscript is probably an error. How the reference to choice of colour came to be added to the Welsh version is unclear, but, along with the aberration in the Corpus manuscript, it might be indicative of a damaged or misread exemplar further down the line. Compare the equivalent passage in the RB-B version, col. 504: *Ac yno y megir dynyon a liwir or lliỏ y mynnont*. The presence of this reference in Version B might indicate that the Corpus reading is not unique and that a similar reference was present in manuscripts of the 1123 version not used by Flint in her edition (or not surviving). Another possibility is that this may reflect cross-contamination between the Welsh translations (another indication of this would be the conclusion of the text at Chapter 81[76] discussed above, pp. 18–20).

adar y griffyt The word can be used for crane or griffin; see *GPC*, s.v. *griff, grifft*. For a discussion, see *DB*, p. 119. There is a parallel to the legend of the battle of Pygmies and cranes in the designation of Arimaspi as traditional enemies of the griffins. See Herodotus, *Histories*, III. 16 and IV. 13, 27; this was transmitted into Latin by Pliny, *Natural History*, VII. 2. See Rudolf Wittkower, 'Marvels of the East. A Study in the History of Monsters', *Journal of the Warburg and Courtauld Institutes*, 5 (1942), 159–97 (p. 160).

yr rỏng y mynyd hỏnnỏ a'r mor y dỏyrein The presence of two definite articles in *a'r mor y dỏyrein* is problematic. *DB*, p. 27, emends by removing the first article. Note, however, that there are other examples of double articles in Middle Welsh; for a discussion see Séamas Ó Gealbháin, 'The Double Article and Related Features of Genitive Syntax in Old Irish and Middle Welsh', *Celtica*, 22 (1991), 119–44. I am grateful to Simon Rodway for this reference. The original Latin text has no reference to the east: *Inter quem et mare Gog and Magog* (*IM*, ed. by Flint, p. 53). The B version also has no reference to the east; see the RB text: *Ac y rỏng hỏnnỏ a mor Got a Magot* (RB-B version, f. 122r, col. 504). It may be that the reference to the east was added as an afterthought, and that the phrase was intended to be *ar y dwyrein*, 'to the east': 'between that mountain and the sea, to the east...', where the second *ar* was omitted due to eye-skip.

Gog, a Magog For Gog and Magog in the medieval tradition, see, for instance, Andrew Gow, 'Gog and Magog on Mappaemundi and Early Printed World Maps: Orientalizing Ethnography in the Apocalyptic Tradition', *Journal of Early Modern History*, 2 (1998), 61–88, and Victor I. Scherb, 'Assimilating Giants: The Appropriation of Gog and Magog in Medieval and Early Modern England', *Journal of Medieval and Early Modern Studies*, 32 (2002), 59–84.

ac eu hynyssed hyt yr awyr. The *IM* has *quorum silvę tangent ęthera*, 'of which the forests touch the air'. Compare the reading of RB-B (f. 122r, col. 504), *a choedyd y rei hynny a ant y'r awyr*, 'and the forests of those ones go to the air'.

Pigeneos This marks the beginning of the discussion of monstrous races, treated in this and the following chapters; the notion of humanoid 'monsters' excited a particular fascination in medieval audiences, and descriptions of them, mostly based on Augustine and Isidore, circulated widely in a range of texts. The description here clearly follows Augustine's *De Civitate Dei* (*On the City of God*), XVI. 8, rather than Isidore (for corresponding information, cf. *Etymologies*, XI. iii). In this and the following chapter, the text refers to a number of races that are mentioned frequently in the medieval tradition, including the Brahmins (*Bragmanos*), the Arimaspians (*Arismapi*), the Cyclopes (*Siclopes*), and the Blemmyes (*Lemenii*). For a general overview of the medieval tradition, see John Block Friedman, *The Monstrous Races in Medieval Art and Thought* (Cambridge, MA: Harvard University Press, 1981), Asa Simon Mittman, *Maps and Monsters in Medieval England* (New York and London: Routledge, 2006), and essays in *The Ashgate Research Companion to Monsters and the Monstrous*, ed. by Mittman and Dendle. For more on the reflection of this tradition in an insular context, see Michael Clarke, 'The Lore of the Monstrous Races in the Developing Text of the Irish *Sex Aetates Mundi*', *CMCS*, 63 (2012), 15–49, and Erich Poppe, 'Exotic and Monstrous Races in the *Leabhar Breac*'s Gospel History and the Transmission of Arcane Knowledge to Medieval Ireland', in *Lochlann. Festskrift til Jan Erik Rekdal på 60-Årsdagen*, ed. by Cathinka Hambro and Lars Ivar Widerøe (Oslo: Hermes Academic Publishing, 2013), pp. 39–56. For more on Brahmins in the medieval European tradition, see Thomas Hahn, 'The Indian Tradition in Western Medieval Intellectual History', *Viator*, 9 (1978), 213–34.

Cyfut = cubit. A frequently occurring (variously spelt, e.g. *cupyt*) term of measurement used in *Delw y Byd*. For the term and spelling variants, see the Glossary and also *GPC*, s.v. *cufydd*. The cubit, like most ancient and medieval terms of measurement, is based on the length of a human body part, in this case the arm from the tip of the middle finger to the elbow. The foundation of the length on human anatomy led naturally to a degree of variation between the various cubits used in different areas and at different times, making it difficult to calculate a contemporary approximation. For more on ancient values of measurement, see Roche, *The Mathematics of Measurement*, pp. 23–24.

A phann loscer y mynyded y ffo y seirff The narrative concerning the origins of black pepper had a wide circulation in the Middle Ages. The *IM* contains only a short version of this story; cf. *IM*, ed. by Flint, p. 53. The full version explains that in order to reach the pepper, which is surrounded by snakes, the fields are set on fire, burning the snakes but also burning the pepper, which therefore acquires its black and crumpled appearance. For a discussion and further references, see Paul Freedman, 'Spices and Late-Medieval European Ideas of Scarcity and Value', *Speculum*, 80 (2005), 1209–27.

griffonnyeit See note to *adar y griffyt* above.

bobloed a elwir 'Agroctas' a Bragmanos The former is in inverted commas because it is framed by punctuation marks, suggesting that the word is being marked out as the object of *elwir*, while Bragmanos are probably not. For more on the wonderous races, see note to *Pigeneos* above.

a ant yn y tan y eu llosgi This appears to be a reference to the practice of the *sati*. For a discussion of the European representation of India and the focus on this custom, see, for instance, for a later period, Joan-Pau Rubiés, *Travel and Ethnology in the Renaissance: South India through European Eyes, 1250–1625* (Cambridge: Cambridge University Press, 2000), pp. 108, 116.

ac ennwir y barnant ar ny wnel uelly. For a discussion of cannibalism see Simon Rodway, 'Affectionate Cannibalism and the Blood Drinking Motif in Gaelic Literature', *CMCS*, 74 (2017), 47–65, esp. pp. 61–62, and, in a more general context, Jill Tattersall, 'Anthropophagi and Eaters of Raw Flesh in French Literature of the Crusade Period: Myth, Tradition and Reality', *Medium Aevum*, 57 (1988), 240–53.

11[12]

'Arismapi', a 'Siclopes' For a bibliography, see note to *Pigeneos* above, p. 84.

a thra orffóyssont ar y dayar The *o* in *orffóyssont* is drawn like an *e* in the manuscript. However, impf. 3 pl. *gorffóyssent* would not work in the sentence.

Lemennii These are Blemmyes. For a bibliography, see note to *Pigeneos* above, p. 84.

12[13]

seucocreta For a discussion of this and other creatures mentioned in this chapter, see N. I. Petrovskaia, 'L'image du monde animalier', in *Mondes animaliers au Moyen Age et à la Renaissance — Tierische Welten im Mittelalter und in der Renaissance*, ed. by Danielle Buschinger, Florent Gabaude, Marie-Geneviève Grossel, Jürgen Kühnel, and Mathieu Olivier (Amiens: Presses du Centre d'Etudes Médiévales de Picardie, 2016), pp. 313–26.

malwot is a word otherwise used for snails; according to the *GPC* this is the earliest instance of the word being used to designate a tortoise (Latin *testudo*, pl. *testudines*; *IM*, ed. by Flint, p. 55); *GPC*, s.v. *malwod*. A later hand wrote in English in the margin of the manuscript at this point: 'A Turtle / Sea Tortoise'.

ac oc eu kogyrneu y góneir lluesteu didos y dynyon. At this point in the text, Version B of the Welsh text preserves additional information about magnets. The WB version reads: *Ac yn yr Indie y mae magneten maen gwyn a tyrr haearnn*, 'And in India there is a shining *magnet* stone, which pulls iron' (f. 2r). This derives from the 1123 version of the *IM*, which forms the basis of *Delw y Byd* Version B. The lack of this information in the A version is one of the features that demonstrate its derivation from the shorter, 1110 version of the *IM*.

20[21]

yn y lle y mae Here and elsewhere in the text, the relative phrase *yn y lle*, 'where, in the place where', is used to translate the Latin *in qua*, 'in which'. The same relative phrase is used again in the chapter to translate the Latin *ubi*, 'where'. For a discussion, see Introduction above, p. 35.

Odyna y mae Galathia a ennóit y gan Gallis Here, *Gallis* is a plural; as it is in the original Latin text; *y rei* refers to it (as *quos* does in Latin). In the Latin the text reads: *Huic iungitur Galatia a Gallis dicta, quos Bythinus rex in auxilium evocavit* (*IM*, ed. by Flint, pp. 58–59), 'This connects to Galatia, named from the Gauls, whom king Bythinus summoned for help'.

Ylyon A reference to the poetical name for Troy, Ilium.

Ovyd The poetic works of the Roman poet Publius Ovidius Naso (43 BC–*c.* 17 AD), particularly the *Metamorphoses*, circulated widely in medieval Western Europe. For a discussion of the knowledge of Ovid in Wales, see Paul Russell, *Reading Ovid in Medieval Wales* (Columbus: Ohio State University Press, 2017); for more on Ovid in medieval Europe more generally, see *Ovid in the Middle Ages*, ed. by James G. Clark et al. (Cambridge: Cambridge University Press, 2011).

Clemens bap The reference is probably to Clement I, pope 88–99 AD, also known as St Clement, probably in particular to his martyrdom (by being thrown into the Black Sea) rather than to his birth; see David Farmer, *The Oxford Dictionary of Saints*, 5th edn, revised (Oxford: Oxford University Press, 2011), s.v. Clement.

21[22]

a gauas y henꞶ A second *Europa* is supplied by *DB*, p. 39, to start the new chapter. However, as it stands the sentence functions as a relative clause to *Europa* (at the end of the previous chapter). Although this suggests that the modern chapter divisions adapted by Flint from Latin manuscripts of the *IM* do not correspond to the text structure perceived by the redactors of the Welsh Version A, it should be noted that both the *o* preceding *Europa* and the *a* following it are marked in red, suggesting a desire to mark the word as a turning point in the text.

Auon Tanais The Don river, following classical geography, considered in the medieval period to be the dividing boundary between Europe and Asia; it provided the left-hand branch of the T in the T-O *mappae mundi*; for the textual source, see Isidore, *Etymologies*, XIII. xxi. 24. For discussions of the boundaries of Europe and Asia, see, for instance, W. H. Parker, 'Europe: How Far?', *The Geographical Journal*, 126 (1960), 278–97.

Philadelphia

y gan a Thanais vrenin The *a* preceding Tanais is an error. The lack of space between the two suggests that the scribe may have perceived *Athanais* as the name.

23[24]

Ratispona This is a reference to Regensburg, the city with which Honorius Augustodunensis is associated; see Introduction above, p. 2.

24[25]

a Llychlyn The modern understanding of geography would require a full stop after *Llychlyn*.

Constantinus The reference here is to Constantine the Great (*c.* 272–337), founder of Constantinople. An important figure for medieval European thinkers, Constantine was responsible for ending the persecution of Christians in Rome, and a document composed in the eighth century and associated with the name of Constantine, the *Donatio Consantini*, 'Donation of Constantine', transferred the power over the territory of the Roman empire to the pope; Brian Z. Tamanaha, *On the Rule of Law: History, Politics, Theory* (Cambridge: Cambridge University Press, 2004), pp. 19–20.

27[29]

Pan doeth hõnnõ gyt ac Eneas o Tro This is an allusion to the foundation legend of Rome, which was founded, according to Virgil's *Aeneid*, by the Trojan Aeneas and his followers, survivors of the Fall of Troy. From the mid-twelfth century onward, foundation legends linking lands, cities, and peoples with Aeneas and his followers proliferated in Western Europe. The legend of the foundation of Britain by Brutus, a descendant of Aeneas, present already in the *Historia Brittonum* and promulgated by Geoffrey of Monmouth in his *Historia regum Britanniae*, and transmitted in Welsh in the translations of his work, *Brut y Brenhinedd*, belongs to this type. For the *Brut y Brenhinedd* passages, see *Historical Texts*, ed. by Williams, pp. 1–2. For Honorius's possible sources, see *IM*, ed. by Flint, p. 62 n. 27.

Ffreinc Liõn The original Latin is *Lugdunum*, which is, indeed, Lyons. This is one of the few instances where the Welsh text provides a Welsh equivalent for a place-name, which suggests that the place-name was familiar. Another example is *Gõasgõin*; for a discussion of possible historical connotations, see note to *Gõasgõin* below.

Comae^t a Note that here the correction appears to have been made by a later hand; possibly the one responsible for the note about tortoises above.

Odyna y rõng Rodõm a Liger y mae Gõasgõin. Latin: *...versus occidentem Aquitaniam, ab aquis Rhodano et Ligere dictam* (*IM*, ed. by Flint, p. 62), 'towards the West Aquitaine, named from the waters of Rhône and Loire'. The Welsh *Gõasgõin* means Gascony. The two rivers in question are the

Rhône and the Loire. Gascony and Aquitaine both lay between the two rivers; Gascony was on the border with Spain and Aquitaine lay north of Gascony. This passage appears to reflect re-writing on the part of the Welsh translators (or their exemplar), displaying a familiarity with this region. Such re-writing is unparalleled in other geographical sections of *Delw y Byd*. If the familiarity displayed with the region and the substitution of Gascony for Aquitaine is a Welsh contribution to the text, it may be due to Welsh participation in Edward I's Gascon campaigns in the years 1296 to 1298; see Adam Chapman, *Welsh Soldiers in the Later Middle Ages* (Woodbridge: Boydell Press, 2015), pp. 22, 24–28. If this supposition is correct, this passage would have to have been produced after 1296. The acephalous Version A text of the thirteenth-century Peniarth 17 only starts later, in the section on water, making it impossible to confirm the form of this passage in that manuscript. Version B treats these regions in a much briefer way and does not have the equivalent passage.

Red Book Version B

Agallica Not capitalised in the manuscript but written together. The word separation and lack of syntactic justification for a separate word *a* suggests that this was a misreading or a scribal error for the place-name *Gallica*. The original Latin is *Gallia Belgica* (*IM*, ed. by Flint, p. 62). It seems that the Welsh translators were unable to interpret this place-name.

Eneas Ysgoydwyn The epithet 'White-Shield' is given to Aeneas in both the Welsh and Latin version of the biography of Gruffudd ap Cynan as well as in *Brut y Brenhinedd*, the Welsh translation of *Historia regum Britanniae*. For the relevant section of *Historia Gruffudd ap Cynan*, see *A Medieval Prince of Wales: The Life of Gruffudd ap Cynan*, ed. by D. Simon Evans (Lampeter: Llanerch, 1990), p. 24 and note to line 13 on p. 91, and *Historia Gruffud vab Kenan*, ed. by Evans, §3; for the Latin version, see *Vita*, ed. by Russell, p. 54 (note that the epithet *Ysgwydwyn* is in Welsh in this text, though the rest of the text is in Latin). For relevant extracts from different versions of the Welsh text of *Brut y Brenhinedd*, see Edmund Reiss, 'Welsh Versions of Geoffrey of Monmouth's *Historia*', *Welsh History Review*, 4 (1968), 97–127 (pp. 115–16), and *Historical Texts*, ed. by Williams, p. 1. For a discussion, see *Trioedd Ynys Prydein: The Triads of the Island of Britain*, ed. and trans. by Rachel Bromwich, 3rd edn (Cardiff: University of Wales Press, 2006), pp. 348–49; also Brynley F. Roberts, 'The Treatment of Personal Names in the Early Welsh Versions of *Historia Regum Britanniae*', *BBCS*, 25 (1973), 274–89 (p. 286).

28[30]

odyna y mae yr Yspaen The initial word of the chapter is not capitalised here, as in the manuscript this appears to be a continuation of the last sentence of the previous chapter. The lack of capitalisation in the manuscript suggests that modern chapter divisions, based on manuscripts of the Latin tradition, do not correspond to the structure of the text as perceived by the Welsh redactors.

hyt y mor Below the final line of the column the words *y gorllewin* are written in the hand of the main scribe, surrounded by a red box with additional decorations.

Cartago The reference here is probably to Cartagena, rather than to the North African city of Carthage; see also *IM*, ed. by Flint, p. 62.

29[31] Red Book Version B

anatred This is a plural form of n.f. *neidr* 'snake', with a prosthetic *a*; see *GMW*, p. 12.

29[31] White Book

Britannia, Anglia, Hybernia These three names exemplify the WB tendency to maintain the Latin forms of place-names, while the RB version tends to substitute Welsh equivalents.

.vj. mis The WB is unique among the Welsh versions in maintaining the use of the Roman numerals as in the original Latin.

30[32]

Yr Affric a dechreu yn dôyrein Indus auon. Renders the Latin [*Affrica*] *in oriente Indi fluminis surgit* (*IM*, ed. by Flint, p. 63), 'Africa rises in the east of the Indus stream'. Note that this geographical conceptualization in the Latin appears to have been confusing to at least one medieval scribe, as the Corpus manuscript, which is related to the Welsh texts (as noted in the Introduction above), replaces *Indi* with *Nili*, bringing the Asia/Africa boundary to coincide with the Nile, as it does in the *mappa mundi* preserved in the same manuscript. The river *Indus* is mentioned in Chapter 10[11] above. Note also that Egypt in medieval T-O maps, as well as in this text, is located in Asia.

Syrenaica a gauas y henó y gan Syren urenhin Punctuation before *y gan* removed. In the Latin, the name of the country *Cyrenaica* is derived from a city, in turn derived from the name of a queen credited with building the city; *IM*, ed. by Flint, p. 63. For an overview of the legend, see Claude Calame, 'Structuralism. Narrating the Foundation of a City: The Symbolic Birth of Cyrene', in *Approaches to Greek Myth*, ed. by L. Edmunds (Baltimore and London: Johns Hopkins University Press, 1990), pp. 275–341 (pp. 277–78), and for more on the city and region, see Jane Fejfer, 'Cyrene and the Cyrenaica', in *The Encyclopedia of Ancient History*, ed. by R. S. Bagnall et al. (Oxford: Wiley-Blackwell, 2012).

y gan e pym dinas yssyd y enó The use of *yssyd* here following the adverbial phrase is unexpected. If it were either 3 sg. ind. or rel. a subject directly preceding would be expected. The subject is *Pontapolis* (= Pentapolis). The colon is editorial. The Welsh mirrors the Latin sentence structure very closely here. See also discussion on p. 34 above.

Beremoe Corrected to *Berenice* in *DB*, p. 45. The error in the manuscript represents a misreading of the letterforms of the exemplar. Here, the last letter but one is definitely *o* and the *i* has no dot (the scribe usually marks the *i* with a thin diagonal stroke above the line to distinguish it from other minims).

yn y lle y bu Aóstin yn escop The construction *yn y lle y* introduces a subordinating phrase. For a discussion, see the Introduction above, p. 35. The reference is to the patristic writer and author of *De civitate dei*, St Augustine of Hippo (354–430); for more on Augustine, see *The Cambridge Companion to Augustine*, ed. by E. Stump (Cambridge: Cambridge University Press, 2001).

32[33]

Ethiopia arall There was a degree of confusion surrounding the concept of Ethiopia in medieval geography. It was occasionally, for instance, confused with India, a confusion dating back to the classical world; see, for instance, Friedman, *Monstrous Races*, pp. 8, 15, and Mary Baine Campbell, 'Asia, Africa, Abyssinia: Writing the Land of Prester John', in *Travel Writing, Form, and Empire: The Poetics and Politics of Mobility*, ed. by Julia Kuehn and Paul Smethurst (New York and London: Routledge, 2012), pp. 21–37. The tradition of multiple Ethiopias appears to thus be linked to the tradition of multiple Indias; see also Geraldine Heng, *The Invention of Race in the European Middle Ages* (Cambridge: Cambridge University Press, 2018), pp. 411–12 n. 165. The texts of the B version (RB-B and WB) appear to omit the reference to the second Ethiopia.

Arall yssyd yn y gorllewin For the 1123 addition which is lacking in this text, see Version B and note to the White Book text (*ac yn y rei hynny*) below.

mor, mόyhaf Translating the Latin *maximus oceanus*. Punctuation mark between *mor* and *mόyhaf* suggests that the scribe did not recognize this as a place-name.

Gadeˢ This is Cadiz, in fact in Spain. Compare the references to Carthage and Troy in the discussion of Italian cities in Chapter 26[28] (on Italy); reproduced in Petrovskaia, '*Delw y byd*'. The Latin original reads: *urbs Gades, a Fenicibus constructa, de qua Gaditanum mare dicitur* (*IM*, ed. by Flint, p. 64), 'city of Gades, built by Phoenicians, from which the Gaditanian Sea is called'.

Yno y mae Promethei y gan y rei y kymerth y mynyded y henό. It is possible that *Promethei* is taken as plural here, as seen in the use of *y rei* to mark a relative construction.

astrologia The word is a borrowing from Latin and is used to mean both astronomy and astrology in classical Latin; *GPC*, s.v. *astrologia*; the distinction was first introduced in medieval Latin; see Lewis and Short, *Latin Dictionary*, s.v. *astrologia*. For knowledge of astrology in medieval Wales, see the discussion in Williams, *Fiery Shapes*, pp. 77–92, 109–10.

o honno y dewir hyt ar y nef. In the Latin, *Vnde et cęlum sustinere dicitur* (*IM*, ed. by Flint, p. 64), 'Whence it is said to sustain the heavens', refers back to Mount Atlas. However, in the Welsh text *o honno* appears to refer back to *astrologia*, as it cannot refer back to *mynyded* (pl. < m. *mynyd*). If *astrologia* is understood here in the broad sense of astronomical knowledge, the reference to 'coming as far as heaven' might be interpreted in the context of *Delw y Byd* itself as the ability to discuss astronomical matter (compare the use of *kerdet*, 'walk', to indicate coverage of a topic in the following sentence).

32[33] Red Book Version B

hanner dyd Rendering the Latin *meridies*, here meaning 'south'. As with the Latin *meridies*, the Welsh term can be used for both 'south' and 'midday'; *GPC*, s.v. *hanner dydd*; Lewis and Short, *Latin Dictionary*, s.v. *meridies*. Compare the RB-A reading *deheu*, 'south'.

Ac yn yr rei hynny y mae ffynnaόn kyn oeret y dyd ac na eill neb y hyuet a chyn bryttet y nos ac na ellir mynet yn y chyόyl. The Latin sentence is *Apud quos est fons tam frigidus diebus ut non bibatur, tam fervidus noctibus ut non tangatur*, 'Near which is a spring so cold in daytime as not to be drunk, so boiling in the night as not to be touched'; this is an addition made to the 1123 version of the text; see *IM*, ed. by Flint, pp. 37, 64. Compare the shorter text of the RB-A version, which is based on the 1110 version of the *IM*. The construction *kyn... ac* in Welsh is equative 'so... as'; see *GMW*, p. 41.

Mor Maϭr a dywedir y vot yn brydyon In this phrase, *brydyon* is the plural form of adj. *brwd*, 'boiling' (*GPC*, s.v. *brwd*); it is unclear why it is in plural form both here and in the White Book text, given that the referent is *Mor Mawr* (Great Sea). One possibility is that this may be an error for *brydyoc*, 'the boiling one', rendering the phrase as 'which is called the boiling one'.

32[33] White Book

keluydyt o'r ser Compare the RB-A reading, *Astrologia*. This appears to be an example of the Welsh translators of Version B interpreting the text, providing a Welsh language equivalent for the Latin word.

33[34]

Y Mor Groec y mae Cyprys 'In the Greek Sea is Cyprus'. Note lack of repetition of the grapheme in this text where the preposition *yn*, 'in', is followed by a word starting with *m-* (and would normally be represented as *ym*). This is also the case in Peniarth 17 (e.g. 86[81] *Y Mars* and *y Mercurius*). The reference is to the Mediterranean; see note to 7[7] *Mor Groec* above, p. 81.

Nawn₁ ynys Seint Daniel The Latin has *Naxon insula Dionysii*, 'Naxos island of Dyonisus'; the name of the island is rendered as *Naron* in some of the Latin manuscripts, suggesting a certain confusion regarding that place-name; *IM*, ed. by Flint, p. 65. *DB*, p. 47, gives *Naiwn* for *Nawn*.

Ynys Varon, yno y byd y marmor gϭynaf, a maen sardinus This translates the Latin *marmor candidissimum quod Parium dicitur, et Sardium lapidem*, 'the whitest marble, which is called Parian, and Sardian stone'. The reference is to the Greek island of Paros (part of the Cyclades group in the Aegean Sea) and to Parian marble, produced in the classical period on that island; Honorius's source for this is Isidore, *Etymologies*, XIV. vi. 28. The stone being referred to as *maen sardinus* is *sardius*. Honorius's source of the reference to *sardius* is Isidore, *Etymologies*, XIV. vi. 29 and XVI. viii. 2, but the stone is also referred to in the description of New Jerusalem in Revelations 21. 20.

In the Welsh text, the -*us* is abbreviated, following three minims of which the first is marked as an *i*. It is possible that the -*us* abbreviation represents a misreading by a previous scribe or translator of a suspension mark over *u*, which would correspond to the accusative ending of the word in the Latin text: *sardium* (*IM*, ed. by Flint, p. 65).

Pitagoras The reference is to the ancient philosopher and mathematician Pythagoras of Samos (570 BC–495 BC). For Pythagorean thought in the Middle Ages, see Christiane L. Joost-Gaugier, *Measuring Heaven: Pythagoras and His Influence on Thought and Art in Antiquity and the Middle Ages* (Ithaca and London: Cornell University Press, 2006), pp. 67–76 and 116–33.

Sibilla The reference is to the Samian Sibyl. Throughout the medieval period the Sibyl remained one of the major figures associated with prophecy; for more on medieval Welsh references to the Sibyl, see Marged Haycock, 'Sy abl fodd, Sibli fain: Sibyl in medieval Wales', in *Heroic Poets and Poetic Heroes in Celtic Tradition*, ed. by J. F. Nagy and L. E. Jones (Dublin: Four Courts Press, 2005), pp. 115–30, and Nely van Seventer, 'The Translation of the *Sybilla Tiburtina* into Middle Welsh', in *'Y Geissaw Chwedleu': Proceedings of the 7th International Colloquium of Societas Celto-Slavica*, ed. by Aled Llion Jones and Maxim Fomin (Bangor: Bangor University, 2018), pp. 109–18. I am grateful to Simon Rodway for the latter reference. The immediate source for the reference to both Pythagoras and the Sibyl is Isidore, *Etymologies*, XIV. vi. 31.

34[35] <35[36]>

Ny enir yno This appears to follow on from the previous chapter, the sentence referring back to the Stecades. It is worth noting that in the manuscript there is no capital or red initial following the punctuation after *Stecades*. It therefore seems that the copyist perceived this as a continuation of the previous sentence.

solifuga The term *solifuga* is in the Latin; *IM*, ed. by Flint, p. 65. The origin of the reference is Isidore, *Etymologies*, XII. iii. 4. For a discussion of the various references to the creature, which is related to both spiders and scorpions, see Kenneth F. Kitchell, Jr, *Animals in the Ancient World from A to Z* (London and New York: Routledge, 2014), pp. 174–75. In contemporary use, the term designates an animal order (also known as sun spiders) within the arachnid class.

apiascon There is a distinct space between *a* and *piascon* in the manuscript, suggesting that the word was unfamiliar to the scribe, but since this is a single recognisable word the space has been removed in this edition. This is the plant which was also known as *apium rusticum* and associated with Sardinia throughout the medieval tradition; for a discussion, see Vincenzo Ortoleva, 'The Meaning and Etymology of the Adjective *Apiosus*', in *'Greek' and Roman in Latin Medical Texts*, ed. by Brigitte Maire (Leiden: Brill, 2014), pp. 259–88.

ac a See note to Chapter 9[10] *Gyon ac a elwir 'Nil'*, above. Another example of this construction in this chapter is *ac a sodes a hi a'e phobyl*.

pan yttoed Plato yndi yn yscriuennu. The reference is to the legend of Atlantis, described by Plato in the dialogues *Timaeus* and *Critias*; for a discussion and references, see Christopher Gill, 'Plato's Atlantis Story and the Birth of Fiction', *Philosophy and Literature*, 3 (1979), 64–78. The formulation in Welsh, however, seems to imply that Plato went down with Atlantis, creating a story parallel to that of Pliny the Elder and the eruption of Vesuvius.

Ac a oed vȯy y medynt no'r Affric, ac Europa. This is problematic. It is possible that a noun qualified by *mȯy* is missing here. The Latin reads: *quę Africam et Europam vicit sua magnitudine* (*IM*, ed. by Flint, p. 66), 'which surpassed Africa and Europe in its magnitude'. It is possible that *medynt*, 'they said' (3 sg. impf.), is an error for a word signifying magnitude, e.g. *meint* (see *GPC*, s.v. *maint*), or for *meddiant*, 'power' (which would have been *medyant* in the manuscript), reading 'and whose power is greater than Africa or Europe'. Alternatively, this could represent a gapped construction, to be interpreted as 'and [the large island = *ynys uaȯr*] was larger, they said, than Africa and Europe'.

prenn ebenus Ebony. This is one of the two early instances of this word listed in *GPC*, and in both cases it appears in translated texts very close to the original Latin and can therefore be considered to be a loan-word; *GPC*, s.v. *ebenus*; cf. Latin original: *In hac est lignum ebenum* (*IM*, ed. by Flint, p. 66), 'in that [island = Meroe] there is ebony wood'.

Cyrene See note to 30[32] *Syrenaica a gauas y henȯ y gan Syren urenhin* on p. 91 above.

Sef oed hynny colledic o bop ffrȯyth a thegȯch o neb ryȯ dim, pell y lles o neb ryȯ dayar, anednybygedic y baȯp or darffei o neb ryȯ damwein y chaffel, odyna pan geffit ny cheffit yno dim... The interpretation of this sentence is problematic and depends somewhat on punctuation. It is possible to read *pell* as expressing superiority ('surpassing') rather than distance. The Welsh diverges somewhat from the Latin in its suggestion that the island is 'deprived of all fruit and beauty of any kind at all'. Thus the Welsh translator appears to have misinterpreted the description of the island as relating to its nature rather than to its inaccessibility. The Latin reads: *Est quedam oceani insula dicta Perdita, amenitate et fertilitate omnium rerum praecunctis terris longe praestantissima, hominibus incognita, quae aliquando casu inventa, postea quaesita non est reperta, et ideo dicitur Perdita* (*IM*, ed. by Flint, p. 66), 'There is a certain island of the ocean called Perdita, surpassing by far all lands in luxury and fruitfulness of all things, unknown to man, which if found at any time by chance, afterwards looked for is not found again, and for that reason called Perdita'.

anednybygedic Sic. Corr.: *anednybydedig*. It is possible the scribe did not understand the sentence or merely did not recognize the word, since adjectives ending -*edic*, typical for translated texts, may not necessarily have been part of normal vocabulary.

yno y dywedit dyuot Brendanus. This is a reference to the legend of St Brendan and his sea travels. In Wales, he is referred to in Rhygyfarch's *Life of St David*; see Richard Sharpe and John Reuben Davies, ed. and trans., 'Rhygyfarch's *Life* of St David', in *St David of Wales: Cult, Church and Nation*, ed. by J. Wynn Evans and Jonathan M. Wooding (Woodbridge: Boydell Press, 2007), pp. 107–55 (pp. 134–37). A version of the legend, which was very popular in medieval Europe, was incorporated into the French translation of the Latin text; see Gossouin de Metz, *L'Image du Monde*. For more on St Brendan, see *The Voyage of Saint Brendan: Representative Versions of the Legend in English Translation*, ed. by W. R. J. Barron and Glyn S. Burgess (Exeter: University of Exeter Press, 2002); for further reading, see Glyn S. Burgess and Clara Strijbosch, *The Legend of St Brendan: A Critical Bibliography* (Dublin: Royal Irish Academy, 2000).

36[37]

Val hynny y mae uffern ym perued y dayar. This should be seen as the second part of the construction beginning with *megys y*: 'just as the Earth is in the centre of the air, so is Hell in the centre of the Earth'. This is an example of the medieval punctuation differing from modern conventions: rather than emphasising connections and the relationship between grammatical sense units, it serves to demarcate breathing pauses for reading the text out loud.

37[37]

Tartarus heuyt, y gelwir For more on Tartarus and Erebus, and the medieval conceptualisation of the regions of Hell, see, for instance, Hans-Werner Goetz, *Gott und die Welt. Religiöse Vorstellungen des frühen und hohen Mittelalters* (Berlin: De Gruyter, 2012), pp. 112–13.

acheronta Acheron, Styx, and Phlegethon are three of the mythological rivers of the underworld. The rivers Acheron and Phlegethon are mentioned in Virgil's *Aeneid*, VI. 107, 295, and 551.

38[38]

Y Dόfyr yό yr eil defnyd annyanaόl The text here moves to discuss the second of the four elements: water. The transition is marked by a larger, coloured initial *y* preceding the coloured initial *d*.

Yr anodun a elwir yn eigaόn yn y lle ny chaffer beis. Compare the Latin: *Huius inmensa profunditas dicitur abissus, quasi abest fundus* (*IM*, ed. by Flint, p. 68), 'Of which the immense depth is called abyss, since it lacks (= *abest*) a bottom'. The sound similarity-based etymology appears to be lost in the Welsh, although it could be argued that *beis* represents a continuation of the play on the words *abissus* (= abyss) and *abest* (= lacks); for more on this, see Petrovskaia, 'La disparition de *quasi*'.

Nyt vnrym hagen y'r anodun ny bo gwaelaόt idaό. This sentence is problematic and confused. It could be rendered as 'It is not the only meaning/virtue however for the abyss [that] there is no bottom to it', where *grym* refers to either the characteristics of the ocean or to the meaning of the word. The Latin reads: *Huius inmensa profunditas dicitur abissus, quasi abest fundus. Habet tamen fundum, quamvis nimis profundum* (*IM*, ed. by Flint, p. 68), 'Of which the immense depth is called abyss, since it lacks a bottom. It nevertheless has a bottom, albeit exceedingly deep'.

40[40]

Gόregys in Welsh is girdle, belt, or zone (of the Earth), but here the Latin is *aestus*, n.m. 'agitation, boiling' or more specifically 'sea tide'; cf. *IM*, ed. by Flint, p. 68. The word in the Latin may have been misread as *vestis*, 'clothing', by a translator; Lewis and Short, *A Latin Dictionary*, s.v. *vestis*. Alternatively, the reading may represent an erroneous emendation of *gwres* (n.m. 'heat') to *gwregis* by the scribe copying the Welsh text.

Pan y tynno... The *y* is the syllabic form of the infixed pronoun, used with *pan;* see *GMW*, p. 56.

52[47]

Eissoes deu ryϬ annyan yϬ: vn ʸ dϬfyr hallt a melys. The sentence as it stands does not work in Welsh. It represents an attempt to render the Latin: *Dicitur tamen quod aquę natura sit duplex, scilicet salsa et dulcis* (*IM*, ed. by Flint, p. 71), 'It is said nevertheless that the nature of water is twofold, namely salty and sweet'. It may be that the *vn* represents a misread *yu*, which itself would be the result of a scribal error repeating *yϬ*, where the original sentence had the meaning 'Nevertheless two kinds of nature is water, salty and sweet'. It is possible that *vn* might represent a (partially) incorporated gloss, where the two natures were numbered and a scribe accidentally incorporated the first of the two numbers into the text. Alternatively, *vn* may represent an error for *yn*: 'Nevertheless here are two kinds of nature in the water, salt and sweet'. I am grateful to Simon Rodway and Peter Schrijver for these suggestions.

DuϬ a duc pedeir auon... This is not marked as a quotation in *DB*, but is introduced by the phrase *mal yr ysgriuennir*, 'as is written', which suggests the introduction of a quotation or at least a reference; cf. *IM*, ed. by Flint, p. 71. It is a reference to Genesis 2. 10–14, a passage describing the rivers of Paradise.

Ac y'r un ffynnaϬn y baradϬys yd ant dracheuyn, a chyt [?] *yr awedϬr, na chymysc a dϬfyr y moroed, namyn llithraϬ o'r dϬfyr yscaϬn ar watʳthaf y trϬm, ac ymchoelut yna y eu fford dirgel* This sentence translates the Latin passage *Que licet universa mare influat, amaris tamen aquis non commiscetur, sed ut puta levis super graves aquas labitur, et in occultum cursum suum revertitur* (*IM*, ed. by Flint, p. 71), 'Which although it flows into the universal sea, nevertheless does not mix with the salty waters, but as pure light glides above heavy waters, and returns to its hidden course'. Note that *y baradϬys* is added above the line in what appears to be the main scribe's hand. It may represent a gloss.

56[51]

er adar enteu a ehedant enteu Both instances of *enteu* appear to represent adverbial use, 'therefore, then', although the proximity appears unusual.

ual y cocodrilli a'r moelronyeit... The term *cocodrilli* is an example of terms retained by the Welsh translators in their original Latin form (nominative plural in this instance). The word *moelronyeit* translates the Latin *ipopotami* (in the Corpus manuscript; see *IM*, ed. by Flint, p. 72 n. 56(6)). In Welsh *moelronyeit* is used miscellaneously for sea/water creatures, with meaning ranging from dolphin, to porpoise, to hippopotamus (as here); *moelron* appears to derive from a combination of *moel*, 'bare' + *rhon/rhawn*, 'tail' or 'spear' (cf. Ir. *rón*, 'seal'); see *GPC*, s.v. *moelrhon*. The information in the *IM* on the capacity of crocodiles and hippopotami to survive both on land and in water is derived from Isidore, *Etymologies*, XII. vi. 3, 19, 21.

Paham... Introduces a direct question, and *urth uot* introduces the answer, without a finite verb. The question mark here is editorial. The question-answer format is frequently encountered in medieval scholarly literature and is adapted by Honorius in the *Elucidarium* and also in the introductory letters to the *IM*, reproduced in the Appendix; see also above, p. 46.

57[52]

ac a agorant yr awyr. The implied subject is probably the waves; in Latin it is clouds, but *wybyr* is singular.

E dyfred ry gerdassam, esgynnôn weithyon e'r awyr. This sentence marks the transition between discussions of the elements. Whilst often described as a bi-partite 'geographical and astronomical' text, Book I of the *IM* is in fact structured around the four elements, moving from Earth (generally described as the geographical section) through Water, Air, and Fire (dealing largely with astronomy). As we have seen, the fragment in Peniarth 17 begins part-way through the section on Water. The RB reads: *O'r dôvyr y traethassam. Weithon y traethôn o'r awyr.*

58[53]

O'r awyr This is followed by a large coloured initial A. The division between sections dealing with each of the four elements therefore appears to be maintained. Compare the beginning of the section on Fire (72[67]) and on Water in the RB (38[38]).

ac o hônnô y kemerant corforoed pan ymdangossont y denyon. The idea that the lower regions of the air are populated by evil spirits is present in both classical and in early Christian writings (supported by biblical passages). They also occur in an insular context, including Bede and Columba's

Altus prosator, for instance. For patristic reference, see Ephesians 2. 2, Augustine, *De Civitate Dei*, VIII. 22 and 'Concerning the Nature of Good'; for the latter, see *Nicene and Post-Nicene Fathers* IV, *St. Augustine*, ed. by Philip Shaff (New York, 2007), p. 358. For a discussion of *Altus prosator*, see Jane Stevenson, 'Altus Prosator', *Celtica*, 23 (1999), 326–68; for the text and translation see Clemens Blume, 'Hymnodia Hiberno-celtica', *Analecta Hymnica Medii Aevi*, 51 (1908) 257–365 (pp. 271–83), and *The Triumph Tree: Scotland's Earliest Poetry AD 550–1350*, trans. by Thomas Owen Clancy (Edinburgh: Canongate Books, 1998), pp. 95–99; for further discussion, see also Valery V. Petroff, 'Eriugena on the Spiritual Body', *American Catholic Philosophical Quarterly*, 79 (2005), 557–610 (pp. 600–02). Flint refers also to the condemnation of these theories in the Pseudo-Byrtferth's glosses on Bede's *De natura rerum*; for references, see *IM*, ed. by Flint, p. 72 n. 58.

59[54]

nyt dim amgen a chuythat awyr Probably to be understood as 'not any different from the blowing of air' or 'namely the blowing of air'; for more on *nyt amgen* see *GMW*, p. 44.

60[55]

Kentaf or petwar prifwynt... The theory of the twelve winds as described here is based on Isidore of Seville's *De natura rerum*, XXVII. 218–19.

Aquilo a Boreas Cf. the Latin: *Eius sinister Aquilo, qui et Boreas* (*IM*, ed. by Flint, p. 73), 'To its left Aquilo, also known as Boreas'. The construction is used in Welsh several times in this passage to render the Latin *qui et*, and it is unclear whether the Welsh translator was aware that this represented two names for the same wind, particularly since a different method is used later on to render the same Latin construction (*Fauonius heuyt yu*).

Subsolanus, gôynt y gollewin... The reference to Subsolanus (note retention of the Latin nominative singular ending) as the wind of the west is an error. While the order in which the winds are named in this manuscript is the same as elsewhere, west and east appear to be transposed (see Figure 7).

y assw enteu wyber a ôac. A wind name is missing here; compare the Latin: *Eius sinister Eurus, nubes generans* (*IM*, ed. by Flint, p. 73), 'To its left Eurus, generating clouds'. Eurus is also missing in the corresponding section of RB-A (see corresponding note below). A reference to Eurus replaces Nothus at a later point in the RB-A text in the context of the discussion of what in the original Latin is *Euroauster*.

Euroauster The two wind names are produced with a space between them in the manuscript, suggesting that the translators or scribes might not have been aware that this represents a single wind, Euroauster; cf. *IM*, ed. by Flint, p. 73. The RB-A reading, by contrast, separates Euro and Aɾster into two winds and presents a description not devoid of internal logic.

O'r tu assw idaɓ enteu Chorus ac Argestes, wybyr a uac en e dwyrein ac eglurder en er India. Cf. the Latin *Eius sinister Chorus, qui et Argestes, in oriente nubile, in India faciens serena* (*IM*, ed. by Flint, p. 73), 'To its left Chorus also known as Argestes, in the east makes clouds, in India clear weather'. The idea expressed here is that the WNW wind Chorus, blowing eastwards, clears clouds from India by pushing them further east.

<61[56]>

ac eno a dewheir yn gestyll ac wybyr. The Welsh renders the Latin *quę conglobatę in nubes densatur* (*IM*, ed. by Flint, p. 73), 'which accumulated condense into clouds'. The use of *yn gestyll* is odd here and may be due to a mis-copying. This instance is the only one provided by *GPC* for the use of *castell*, 'castle', for 'cloud'; see *GPC*, s.v. *castell*. However, it is worth noting that as a reference to cumulonimbus clouds the metaphor of a castle seems particularly apt. I am grateful to Paul Russell for this idea.

Sef yu wybyr llongeu y cawadeu. This represents an attempt to translate the sound-based etymology given in the Latin original: *Dicuntur autem nubes, quasi nimborum naves* (*IM*, ed. by Flint, p. 73), which relies on the sound similarity between *nubes* and *nimborum naves*, 'they are called clouds, as it were ships of clouds/rainstorms', lost in the Welsh translation. For a more detailed discussion of the Welsh translators' treatment of etymological formulae, see Petrovskaia, 'La disparition du *quasi*'.

60[55] Red Book Version A

Deheuwynt hɔnnɔ. In the first part of this passage *deheuwynt* is used to designate the wind to the right (Lat: *dexter*) of the previous wind and *gogledwynt* the wind to the left (Lat. *sinister*) of the previous. The use of *deheu* and *assw* in the Peniarth 17 text suggests that this may be due to an interpretation of a Welsh exemplar's use of *deheu* for 'right' as 'south' and the subsequent correction of *assw* to *gogled*. It is worth noting that *gogled* is derived from *go-* + *cled* (= adj/n.f. 'left'/'left side'); *GPC*, s.v. *gogledd*. Note also that *asseu* is used further on for the unnamed companion wind of Subsolanus and for the left-hand winds thereafter.

Deheuwynt h6nn6 y6 Uulte^{u}rnus, pob peth a sycha The reference to the other name of this wind, Calcias, is missing here; cf. the equivalent passage in Peniarth 17.

a'r g6ynt o'r tu asseu ida6 ac a uac wybyr. There is a wind name missing here, *Eurus*. Compare the reference to Eurus in the Latin: *Eius sinister Eurus, nubes generans* (*IM*, ed. by Flint, p. 73), 'To its left Eurus, generating clouds', and see the note to the corresponding section in Peniarth 17, above. The *ac* may thus be superfluous, but see also discussion of *ac a*, in note to Chapter 9[10] *Eil auon y6 Gyon, ac a elwir 'Nil'*, above, p. 82. Note also the use of *asseu* rather than *gogled* here to designate the left wind.

Ac ar y deheu y mae a^{e}6ro. A6ster g6ressa6c y6, Euro ynteu g6ynt ardymher^{u}s y6. The Welsh has re-organised the winds here somewhat, possibly due to confusion regarding Auster, Eurus, and Euroauster. Compare the Latin: *Huius dexter Euroauster, calidus. Eius sinister Euronothus, temperatus* (*IM*, ed. by Flint, p. 73), 'To the right of which Euroauster, hot. To its left Euronothus, temperate'. See the diagrams in Figure 7 above.

m6yhaf The manuscript's *m6yha6^{y}haf* may be the result of accidental repetition, *m6y[ha6^y]haf*, made in the course of correcting 6; cf. the Latin: *Australes venti faciunt maiores tempestates in mari* (*IM*, ed. by Flint, p. 73), 'Southern winds make greatest tempests in the seas'.

y asseu y6 Chorus, yn y d6yrein y mae wybyr, ac eglurder yn yr India. See note to the equivalent passage in Peniarth 17 above.

60[55] *Red Book Version B*

Awster g6ynt r6^{y}medic y6 The use of the term *r6ymedic*, 'bound' or 'chained', is problematic, as the term should be translating the Latin *calidus*, 'hot'; it is possible that either in the process of transmission in the Latin text or in the process of translation into Welsh *calidus* was misread as *catenatus*, 'chained' (Lewis and Short, *A Latin Dictionary*, s.v. *catenatus*). *R6ymedic* is also the reading in the Rawlinson text.

a lleiaf a wna góynt y de^m hestyl yn y mor. There appears to be an error here. According to the Latin text, the southern winds increase tempests in the sea; *IM*, ed. by Flint, p. 73. A similar error to the RB-B reading is attested in the WB and Rawlinson B 467, where the identification of the winds as southern is, however, retained. The sentence is grammatically problematic in all versions.

A deheu ^yv. This appears to be a fragment of a phrase, unrelated to either the previous or the subsequent sentence. The feature is shared by Rawlinson B 467. It is therefore possible that the problem lay in the exemplar shared by the two manuscripts (since it is not attested in the WB). It may have originally arisen because of a redactor's having become habituated to the *deheu/asseu* alternation of the winds, since the previous wind is designated *asseu.* The two following winds are listed in the Latin text as additional to the twelve winds scheme and thus constitute a sudden change in the pattern, which might have taken the redactor unawares.

60[55] White Book

yrr This is *gyrru*, 'banish'; *GPC*, s.v. *gyrraf: gyrru*. The *h* in the manuscript is superfluous. Note that RB-B has *yrr* in the corresponding sentence.

a chrynnant This also appears to be an error, as lenition would be expected here.

60[55] Rawlinson B 467

Kyntaf yó Septentrio ac ef óna oeruel ac óybyr oer This passage shows the close affinity between the Rawlinson and the RB-B texts, which throughout agree against the WB. There is a preverbal particle *a* missing.

tu deheu Arall a elóir Garaus Capitalisation of *Arall* in MS. There seems to have been some difficulty in the exemplar, since the wind name is changed from *Circius* to *Garaus*.

Eure Nothus Spaced out as two words in the manuscript; this refers, however, to the wind Euronothus; cf. *IM*, ed. by Flint, p. 73.

Lleihaa a óna góynt y deheu The pres. ind. 3 sg. form *lleihaa* (< *lleihau*, 'to diminish') is unexpected here. This may be an error for *lleihau* (vn), but compare also the RB-B and WB readings: *lleiaf a óna*. For more on the extra *-a* in *leihaa*, see Rodway, *Dating Medieval Welsh Literature*, p. 50.

A deheu yɣ See note above to *A deheu* [yv] in RB-B.

63[58]

E bwa en er awyr petwarlliwyauc The medieval four-colour theory of the
rainbow, derived by Honorius from Isidore's *De natura rerum*, but also
appearing in Bede's *De natura rerum*, appears to have been ultimately
derived from Empedocles's association of the four colours he identified
as basic with the four elements (Fire, white; Water, black; Air, red; Earth,
yellow-green); for a discussion and further references, see Kirsten Wolf, 'The
Colors of the Rainbow in Snorri's *Edda*', *Maal og Minne*, 1 (2007), 51–62 (pp.
55–56).

67[62] *Red Book A*

echtywynnedigrɣyd 'brightness'. Not listed in *GPC*; from the adjective
echtywynnedig, 'brightening', and the suffix *-rɣyd* used to create nouns
from adjectives; *GPC*, s.v. *-rwydd, -rhwydd*. The adjective *echtywynnedig* is
in turn derived from the verb *echdywynygu*, 'to shine forth', 'gleam'; *GPC*,
s.v. *echdywynygu*, ending in *-edig*, form typical of translation style. Compare
gwynfydedigrwydd, 'beatitude', 'bliss' (*GPC*, s.v. *gwynfydedigrwydd*).

71[66]

Tymestyl awyraul... This chapter is missing from RB-A. The Welsh text of
Peniarth 17 diverges from the original here, which refers to pestilence being
born of overly dry air, rather than to tempests. This may be due to an error
on the part of the translator or an error in the Latin exemplar confusing
pestilentia/tempestas. A misreading based on the context (a discussion of
air, smoke, fire) is possible, particularly given the presence of the same
letters and sounds in both words. This confusion is unlikely to have taken
place after translation, since the Welsh word used for pestilence would have
most likely been *ball, aball*, or *angheu*. The Latinate borrowing *pestilens* is
not attested in Welsh before the sixteenth century; see *GPC*, s.v. *pestilens*.
The stem of *Tymestyl/tymestl*, 'tempest', is derived from the Latin *tempestas*;
also attested in M. Welsh in the form *tympestyl* and *tymhestyl*; see *GPC*, s.v.
tymestl.

72[67]

E tan A larger coloured initial here marks the commencement of the section on
the fourth element, Fire.

pur awyr The phrase translates the Latin *purus aer*, used in the original to explain *ether* (*IM*, ed. by Flint, p. 75); indeed, 'pure air' is the meaning of the Greek αἰθήρ; see Henry George Liddell and Robert Scott, *An Intermediate Greek-English Lexicon* (Oxford: Clarendon Press, 1945), s.v. αἰθήρ. For more on ether, see the note to 1[1] *Odyna yr awyr glan ygkylch yr wybyr*, above.

o hѹnnѹ y kemer engylyon corforoed pan anѹoner ar deneon. There was a great deal of debate in the Middle Ages as to whether the angels had corporeal bodies; for more on the subject, see David Keck, *Angels and Angelology in the Middle Ages* (Oxford: Oxford University Press, 1998), pp. 31–32.

73[68]

y ѹelly y planedeu hyn en erbyn y sygneu. For a discussion of medieval planetary motion theory, see Grant, *Planets*, pp. 315–20, 497–98.

74[69]

Ny leihaa hi heuyt ac ny muyhaa A scientific explanation of the phases of the moon based on the movement of the shadow of the sun.

gerdet holl sygneu zodiacus yn wyth diwyrnaut ar ugeint. For a discussion of medieval ideas regarding the speeds of planetary rotation, see Grant, *Planets*, p. 494.

74[69] Red Book Version A

wyneb yr heul This may be the earliest instance of *wyneb yr heul* for 'rays of the sun/sunlight'; see *GPC*, s.v. *wyneb*. The earliest instance given in *GPC* is from the sixteenth century.

neuatus Here probably *nef*, 'heaven'.

74[69] Red Book Version B

Ac ef a dywedir pei na bei y dѹfyr, kyn oleuhet vydei a'r heul, a mѹy yѹ no'r dayar. May be read as 'And it is said unless there were the water, it would be as bright as the Sun, and it is greater than the Earth'. The second *na bei* appears to be a scribal error. The reference is to the watery nature of the Moon. The final part of the sentence results from a misunderstanding of the original Latin, according to which, were it not for the water it contains, the Moon would burn the Earth far more than the Sun, because it is closer. The original Latin (closely followed by the Version A readings in Peniarth 17 and RB-A) reads: *Dicitur enim si aqua permixta non esset, terram ut sol illustraret, immo ob vicinitate maximo ardore vastaret* (*IM*, ed. by Flint, p. 76), 'It is said [of it] that if water were not mixed with it, it would have illuminated the Earth like the Sun, indeed from [its] proximity would have devastated [the Earth] with greatest heat'.

81[76]

delw o euyd yn dechreu Sadurn The Latin reads: *In exortu illius post .xxx. annos qui imaginem de ẹre fuderit, loqui ut hominem probabit* (*IM*, ed. by Flint, p. 78), 'He who in its rise after thirty years casts an image in copper/bronze/ brass, will have the right to speak like a man'. Since the text explicitly refers to the ascendant after thirty years, it is probable that the reference is to the astrological concept of 'Saturn return', the return of the planet to the exact same place it occupied at a person's birth, which occurs, at a very rough estimate, around thirty years later. For a discussion of this passage, see G. Ruddock, *Dafydd Nanmor* (Caernarfon: Gwasg Pantycelyn, 1992), pp. 77–79.

81[76] Red Book Version A

Crónn yó, a reólyt; yn erbyn y dayar, ac yn erbyn y plannedeu y danaó a gerda This sentence provides an illustrative example of the difference between medieval and modern punctuation practices. Here, the *punctus* is used to mark pairs joined by conjunctions (it is placed in front of *a* and *ac*) in the two clauses, but no punctuation is used to separate the clauses.

Póy bynnac a dineuo deĺ o euyd, hi a dyweit ual dyn. The use of the feminine pronoun *hi* is strange, at variance with both the Peniarth 17 use of the masculine pronoun *ef* and with the original Latin (see note above). From a structural point of view, it should be referring back to *póy bynnac*. However, it is conceivable that it refers back to *deló*, and that the intention is to suggest the image will have magical properties: it will be able to speak like a man.

81[76] White Book

A phwy bynnac a welei deĺ o efuyd The beginning of the chapter on Saturn is missing in the White Book. This appears to be the result of eye-skip. Together with the uncorrected insertion of *ef a allei* too early in the sentence, this suggests scribal carelessness at this point.

85[80]

Y seith gylch henne Each of the seven 'planets' was perceived as running along one of seven concentric spheres. For more on this, see Simek, *Heaven and Earth*, pp. 6–11.

gywydolyaeth The term also occurs in *Gwyrthyeu y Wynvydedic Veir*. For more on heavenly music, see note to 86[81] *gama en e daear*, below.

86[81]

Gama en e daear. This reflects the medieval theory, derived from the Pythagoreans, of the music of the spheres; see W. C. Dampier, *A History of Science and its Relations with Philosophy and Religion*, 4th edn (Cambridge: Cambridge University Press, 1971), pp. 16–18, 127–28. The chapter describes the octave (in the sense that the octave contains seven notes, usually described in contemporary notation by letters of the alphabet from A to G). The lowest note in the musical scale was conventionally marked with a Γ (*gamma*), and the scale itself was known, after its lowest note, as a *gamut*.

87[82]

y byt lleihaf See note to 92[86] 'Anatole', 'Dissis', 'Arcton', 'Messembria', below.

Additional Material from Red Book A 88[83]

O'r dayar hyt att y lloer y maent pymtheg mil a whechant, a phymthec milltir ar hugeint A degree of variation is present in the punctuation of this passage, with a *punctus* sometimes following *y maent* and sometimes not. In one instance, the use of a coloured initial suggests the beginning of a new sentence (*hyt ar Jouen y maent. Seith mil ac 6yth cant*). For the sake of consistency, and to make the passage somewhat easier to follow, where punctuation follows *y maent*, it has been rendered as a colon. The number *phymthec milltir ar hugeint* is presumably an error for *pump milltir ar hugeint*, as the original is twenty-five.

92[86]

Duyrein a dywedir o achaus dwyrau yr heul ohona6, e gollewin enteu a dywedir o achaus diguydau er heul eno. The etymological explanations of the words for 'east' and 'west' given in the Latin are retained here. Cf. *IM*, ed. by Flint, p. 81: *Oriens ab ortu solis, Occidens ab occasu eius dicitur*, 'The East (= *oriens*) is [so] called from the rising (= *ortus*) of the sun, the West (= *occidens*) from the setting of the sun (= *occasus*)'. The sound-link between 'east' (*duyrein*) and the rising (*dwyrau*) of the sun is retained in Welsh, while the link between 'west' (*gollewin*) and the setting (*diguydau*) is lost. Similarly, the link between south and midday is lost in the Welsh (in Latin the words are *meridies*, 'south', and *medi-dies*, 'midday', translated in Welsh as *hanner dyd*). For more on renditions of etymologies in the Welsh translations, see Petrovskaia, 'La disparition du *quasi*'.

yn yaόn appears to be an interpolation and can be interpreted as either preposition + noun or adverbial *yn* + adjective *yaόn*, 'true', 'correct' (*GPC*, s.v. *iawn*).

'*Anatole*', '*Dissis*', '*Arcton*', '*Messembria*', *ac o'r llythyren gentaf o'r petwar henό henne y kynullόt enw Adaf sef yu hόnnό y byt lleiaf* The construction of Adam's name out of the Greek terms for the four parts of the world (*anathole, disis, anathos, mesembria*) is in Honorius's text derived ultimately from Isidore. Honorius also refers to it in his *Elucidarium*, I. 64; see *L'Elucidarium et les lucidaires*, ed. by Yves Lefèvre (Paris: E. de Boccard, 1954), p. 372. This reflects a more widespread tradition: the medieval concept of man as microcosm (including the link with the Greek names for parts of the world), expressed famously by Augustine, among others; see Naomi Reed Kline, *Maps of Medieval Thought: The Hereford Paradigm* (Woodbridge: Boydell Press, 2001), pp. 227–28; for a discussion in a Celtic context, see Hildegard L. C. Tristram, 'Der *homo octipartitus* in der irischen und altenglischen Literatur', *ZCP*, 34 (1975), 119–53. For a discussion of the concept in the context of medieval Welsh literary culture, and further references, see *Legendary Poems*, ed. and trans. by Haycock, pp. 533–34.

93[87]

hόnnό yu y nef uchaf The original reads *superius cęlum dicitur firmamentum* (*IM*, ed. by Flint, p. 82), 'the highest sky is called the firmament'. For discussions of the concepts of sky, heaven, and firmament in the European Middle Ages, see, for instance, *Envisaging Heaven in the Middle Ages*, ed. by Carolyn Muessig and Ad Putter (London and New York: Routledge, 2007), and Reudi Imbach, '*Empyreum*. Scholastische Gedanken über das Paradies', *Deutsches Dante-Jahrbuch*, 83 (2008), 13–37.

94[88]

ac a welir y gennym ni This can be rendered as 'and seen by us'. However, since this is part of one of two parallel constructions, of which the second is an unambiguous genitival relative clause, *ac a* here could also be interpreted as introducing a relative clause: 'and which is seen by us'; see *GMW*, p. 63.

97[91]

Sef yu enѳ y 'kylch zodiacus' y groec 'kylch y sygneu' yu henne yg kymraec. Compare the Latin: *Horum disposition dicitur Grece zodiacus, latine signifer, eo quod fert signa que animalium habent nomina. Zodin enim dicitur animal* (*IM*, ed. by Flint, p. 83), 'Their disposition in Greek is called *zodiacus*, in Latin *signifer*, that which bears signs that have animal names. *Zodin* namely means animal'. The explanation in the original refers to Latin, and is here adapted for a new audience, referring to Welsh. The final sentence is not translated.

98[92]

Maharaen 'ram' or, as here, Aries; also occurs as Aries in *Brut Dingestow*, ed. by Henry Lewis (Cardiff: University of Wales Press, 1942), p. 116. For the origins of the word see *GPC*, s.v. *maharen/myharen*. For the sources of the legends of this and the following signs, see the notes in *IM*, ed. by Flint, pp. 83–91.

109[103]

pan yt oed Antiocus urenhin en fo rac Uenus a Chupido e map In the Latin it is Venus and Cupid who flee: *Cum dii Tipheum gigantem insequentem fugerent in Egyptum, Venus et Cupido filius versi in pisces latuerunt in aquis* (*IM*, ed. by Flint, p. 85), 'When the gods fled in Egypt from the pursuing giant Tipheus, Venus and Cupid [her] son turned into fish hid in the waters'. For the use of *ffoaf*, 'flee', see also Chapter 10[11] above. The change from Tipheus to Antiochus is strange. King Antiochus is mentioned once in the RB-A text of *Delw y Byd*, in Chapter 15[16]: *Yno y mae Antiochia a gauas y henѳ y gan Antiochus vrenhin* 'There is Antioch, which got its name from King Antiochus'. The city of Antiochia was named by one of Alexander the Great's generals, Seleucus, in honour of his father; see *The Hellenistic World from Alexander to the Roman Conquest: A Selection of Ancient Sources in Translation*, ed. by M. M. Austin (Cambridge: Cambridge University Press, 1981), p. 88.

llyngu Cf. *GPC*, s.v. *llyncu*. For more on the variation between *-nk* and *-ng*, which appears to be partly a matter of date and partly a matter of dialect, with earlier manuscripts (up to the early fourteenth century) and northern manuscripts showing a preference for -*ng*, see Thomas Charles-Edwards, '"Mi a dynghaf dynghed" and Related Problems', in *Hispano-Gallo-Brittonica: Essays in Honour of Professor D. Ellis Evans on the Occasion of his Sixty-Fifth Birthday*, ed. by Joseph F. Eska, R. Geraint Gruffydd, and Nicolas Jacobs (Cardiff: University of Wales Press, 1995), pp. 1–15, esp. pp. 6–10.

127[121]

Sagitta Smaller northern constellation, not to be confused with Sagittarius.

Philotetes Whilst primarily associated with Troy, Philoctetes also features in the legend of Hercules, as the only man willing to light Hercules' funeral pyre, thus winning the latter's gift of his bow and arrows. For the rarity of medieval references to Philoctetes, see H. David Brumble, *Classical Myths and Legends in the Middle Ages and Renaissance: A Dictionary of Allegorical Meanings* (London: Greenwood, 1998), p. xii.

128[122]

Idra caur The reference is to the multi-headed monster Hydra, vanquished by Hercules. The description of the monster as *caur*, 'giant', is not in the Latin and appears to be a Welsh addition. For more on Hercules in the Welsh literary context, see Marged Haycock, '"Some Talk of Alexander and Some of Hercules": Three Early Medieval Poems from the Book of Taliesin', *CMCS*, 13 (1987), 7–38.

140[134]

a ladaud Beloropho The myth of Bellerophon is recounted in the *Iliad*, but was probably known in the medieval period through Fulgentius of Ruspe; it was also included in the *Ovide moralisé*. Note that Bellerophon was often confused with Perseus. For discussions, see, for instance, John M. Steadman, 'Perseus upon Pegasus' and Ovid Moralized', *The Review of English Studies*, 9 (1958), 407–10, and M. Shapiro, 'Perseus and Bellerophon in "Orlando Furioso"', *Modern Philology*, 81 (1983), 109–30.

143[136]

Y guregys llaethaul Translating Latin *lactea via*, Milky Way. Given the abundance of mythology-derived information in this text, it is curious that the description of the Milky Way provides a scientific (light of the stars) rather than a mythological explanation; for the transmission of the classical myths related to the Milky Way, see F. Bertola, 'The Milky Way Through the Ages: An Iconographic Journey', in *Cosmology Across Cultures ASP Conference Series 409*, ed. by J. Rubiño-Martín et al. (San Francisco, CA: Astronomical Society of the Pacific, 2009), pp. 237–41.

eu The last two minims are unclear and could be either *u* or *n*; *DB* has *eu*; but Isaac and Rodway, 'Peniarth 17' has *en*. Either option could work here: 'stars pour out in light' or 'stars pour out their light'. The Latin has *fundunt in ea sua lumina*, 'pour into it their lights', suggesting *eu* might be correct. The next chapter has a seemingly parallel construction in *y syr a dineu en flameu*, which, however, renders a different Latin turn of phrase.

145[138]

nef dyfraul This is the first of the three heavens presented in this text (with *nef ysprydaul* and *nef y neuoed*); Honorius's source for the doctrine of the three heavens appears to be Augustine's *De Genesi ad Litteram*, XII. xxxiv. 67 (*IM*, ed. by Flint, p. 92 n.).

147[140]

nef y neuoed translates the Latin *caelum caelorum*; the phrase represents one of the superlative genitive constructions the prominence of which in medieval Latin is probably due to the influence of the Hebrew syntax in the Bible; see J. B. Hofmann, *Lateinische Syntax und Stilistik*, 5th edn, rev. by A. Szantyr, 2 vols (Munich: C. H. Beck, 1965), II, p. 55, §52, sec. γ; Albert Blaise, *Manuel du latin chrétien* (Strasbourg: Le Latin chrétien, 1955), p. 83 §87. In the Bible, the phrase *caelum caelorum* occurs, for instance, in Psalm 113. 24; see Jean Pépin, 'Recherches sur le sens et les origines de l'expression *Caelum Caeli* dans le Livre XII des *Confessions* de S. Augustin', *Archivum latinitatis medii aevi* (1953), 185–274.

NOTES TO *APPENDIX I:*
INTRODUCTORY LETTERS

[Letter of Christianus]

Seith nawn yr Yspryt Glan The 'Seven Gifts of the Holy Spirit' is a patristic
tradition, founded in Paul's letters, particularly 1 Corinthians; see Stephen
S. Smalley, 'Spiritual Gifts and I Corinthians 12–16', *Journal of Biblical
Literature*, 7 (1968), 427–33.

a gỽedy gorffenno oessoed y uuched honn, llaẟenhau... kytwledychu The use of
the verbal nouns here is problematic. The original Latin sentence is also
problematic in its lack of a verb (conventional omission for letters) and
apparent lack of the conventional *salutem* (usually retained). It is possible
that in the *IM* letter the latter is replaced by the lengthy final section of the
sentence, where the sender is wishing his correspondent spiritual health. If
the Welsh translation structurally follows the Latin, then while the subject
of *llaẟenhau* and *kytwledychu* could grammatically be either the author
of the letter or his addressee, this interpretation would suggest it is the
addressee who is to enjoy the seven blessings and the presence of God after
the end of his life.

yr wythuet kytwledychu It is unclear what *eighth* is being referred to. It seems,
in the context of the Welsh text, to refer to an additional eighth blessing, but
in the original Latin *in octava* may well represent *in octava* [*die*], referring
back to the *post septimanam* of the previous sentence: the eighth day, the day
following the completion of the week. *Kytwledychu* is another instance of
the problematic use of vn in this passage (see comment on *laẟenhau* above).
It is unlikely that this means co-rule, given the ecclesiastical context of the
passage. It may derive from *gwleddychaf*, 'to feast', and have the sense of
celebration, or a spiritual sense, a meaning carried both by *gwledd*, 'feast',
and *gwledd*, 'partage in a feast, carouse, revel' (*GPC*, s.v. *gwleddaf*). The use
of the prefix *kyt-* is in any case probably influenced by the *com/n-* in the
original Latin *comtemplari*. Note that the Welsh equivalent of *contemplare*,
cytweld, is not attested in the medieval period; see *GPC*, s.v. *cydwelaf*, 'to
agree' (first attested in the eighteenth century) or 'to see together' (first
attested in the sixteenth century).

ỻrth hynny y gỻelir y mi wybot Here, *wybot* is probably an error for *uy mot*, misunderstood by the scribe; see *DB*, p. 115.

megys y myỻn claỻr. A direct parallel between the verbal description of the world in the *IM* and its visual description in *mappae mundi* is drawn here.

wedy eu gỻneuthur y rom, a ninheu heb adnabot dim onadunt hỻy Here, *y rom* means 'for us', 'for our sake'; this form of the preposition is not in *GMW*, but see Sims-Williams, 'Variation', pp. 33–34, 46. For *ninheu* (*niheu* in MS), see *GPC*, s.v. *ninnau*.

[Letter of Honorius]

tryzor According to *GPC*, W. *trysor* is a borrowing from Old French or Middle English. The *z* seems to reflect the sound of the Middle French word; see *GPC*, s.v. *trysor*; cf. O. Patterson, 'Honor and Shame in Medieval Welsh Society', *Studia Celtica*, 16–17 (1981–1982), 73–102 (p. 84 n. 1).

val y dyweit y diaereb: gỻlanha ar auyr. The proverb in question features in the A text of the RB collection as *Nyt haỻd gỻlana ar yr afyr*; see *Diarhebion*, ed. by Roberts, line 850; see also his note to the line for a comment on the relation with Honorius's text and further references.

val y gỻahanỻyt o'r defnydyeu kyntaf He describes the structure of the text, which is structured according to the four elements; see discussion in the Introduction above, p. 4.

Ac oc an llauur ni a gymerant yn weithret udunt ehun, ac a sathrant argyhoed mal y mo[ch] y'r mererit. The use of prepositional phrase + *a* is problematic. It is possible that a personal pronoun 3 pl. (*wy/wynt*) is missing between *an llauur ni* and *a*. As it stands, the construction mirrors the Latin closely: *atque de labore nostro sibi scientiam usurpant que ut sues margaritas pedibus proculcant* (*IM*, ed. by Flint, p. 48), 'and of our work usurp to themselves the knowledge, which they trample as swine do pearls'. It seems the translator attempted to use the mixed order to emphasise that it is the speaker's work that is being stolen and trampled. The reference ultimately might be to accusations of heresy which Honorius's writings and some of his sources were subject to; for more see Sanford, 'Honorius', esp. p. 399. The image of swine trampling pearls is attested in other Welsh texts: in the translated text *Brut y Brenhinedd* as *a gwarcarey mereryt adan traet e moch* (see Brynley F. Roberts, *Astudiaeth destuonol o'r tri chyfieithiad Cymraeg cynharaf o 'Historia Regum Britanniae' Sieffre o Fynwy, ynghyd ag 'agraffiad' beirniadol o destun Peniarth 44* (unpublished PhD thesis, University of Wales, 1969), p. 99, cited in *GPC*, s.v. *mererid*), and in Prydydd y Moch/Llywarch ap Llywelyn's poem 'Bygໄth gruffut vab kynan' in the Hendregadredd manuscript as *mal heu rac moch meryerid* (*Gwaith Llywarch ap Llywelyn 'Prydydd y Moch'*, ed. by Elin M. Jones and Nerys Ann Jones (Cardiff: University of Wales Press, 1991), 8. 14).

gan ysgaelussaໄ y dynyon kyghoruynnus, ac a ymdodant ehunein, ac a gnoant ehunein o gallon gyghoruynnvs. Use of *gan* with a verbal noun appears to be participial, 'under/with ignoring the jealous people and what they digest and what they chew with malicious heart'. The two verbs *ymdodant* (< *ymdodi*) and *cnoant* (< *cnoi*) occur in medical texts; that they occur together here suggests that *ymdodant* should be interpreted in a bodily sense as 'digest', rather than 'dissolve'; I am grateful to Elena Parina for this suggestion. The word *ymdodant* here translates the Latin *tabescentes* (nominative/accusative plural present active participle) < *tabesco, tabescere*, 'melt, dissolve'.

goruchelyon defnydyon The agreement of plural adjective and plural noun is unusual in Welsh. It is often considered a feature of what is known as 'translator's style'; *Cyfranc*, ed. by Roberts, p. xxviii; see also Introduction above, pp. 33–37. The phrase appears to be an attempt to translate the Latin *ardua molimina*; *ardua* can mean 'high' as well as 'steep' or 'difficult'; *molimen* means 'effort' or 'undertaking' and it is unclear by what process this resulted in translation into Welsh as *defnydyon*.

Ac ໄrth hynny ar dysc y lawer, nyt o amylder llyfreu udunt y gໄneir y llyfyr hໄnn. This corresponds to the Latin *Ad instructionem itaque multorum quibus deest copia librorum, hic libellus edatur* (*IM*, ed. by Flint, p. 49), 'Therefore this little book is produced for the instruction of the many who lack abundant books'. The Welsh might accordingly be interpreted as 'Therefore, on teaching of many [i.e. in order to teach to many], not with a multitude of books to them [i.e. not having many books], was this book made'. This statement reflects the popularising purpose of medieval Welsh translations observed by Brynley Roberts, '*Ystoriaeu Brenhinedd*', p. 217. It reflects the main objective of this book: to make knowledge accessible to all.

GLOSSARY

Note that composite numerals are not given in the glossary. For these, see the entries for the component numerals. The glossary includes Latin terms and forms retained in the text. A question mark ? is used to indicate instances of uncertainty. Words that are present in the text in mutated form are added to the glossary without mutations. For words which occur more than five times in the text only the first five instances are cited for each form. Where not attested in the text, dictionary forms are supplied in modern Welsh spelling, marked with an *. The order of the English alphabet is followed, excepting c/k which have been conflated.

Grammatical Abbreviations

adj.	adjective		neg.	negative
adv.	adverb		num.	numeral
art.	article		obj.	object
conj.	conjunction		part.	particle
cpv.	comparative		pl.	plural
def.	definite		poss.	possessive
dem.	demonstrative		prep.	preposition/ prepositional
eq.	equative		pres.	present
f.	feminine		pret.	preterite
impers.	impersonal		pron.	pronoun
impf.	imperfect		rel.	relative
ind.	indicative		sg.	singular
inf.	infixed		spv.	superlative
ipv.	imperative		subj.	subjunctive
Lat.	Latin		subst.	substantive
m.	masculine		vb	verb
n.	noun		vn	verbal noun

a

a, a(c)[1] conj. *and*: a RB-B <1[1].2>, RB-A 2[2].9, 2[2].11, RB-A 3[3].5, 3[3].8; ac RB-A 1[1].2, 1[1].3, 2[2].2 (twice), 2[2].6; with inf. obj. pron. 3 sg. m. a'e RB-A 3[3].6; with inf. obj. pron. 3 sg. f. RB-A 50[46].1; with inf. poss. pron. 3 sg. m. a'e RB-A 12[13].14, 37[37].7, LC.1; with inf. poss. pron. 3 sg. f. a'e RB-A 37[37].9, Pen.17 74[69].17; with def. art. a'r *and the*: RB-A 1[1].4, RB-B <1[1].3>, (twice), <1[1].5>, (twice), <1[1].6>; a'r conj. + dem. art. *and that*: RB-A LH.8

a, a(c)² prep. *with*: **a** RB-A 2[2].10, 10[11].13, 11[12].1. 11[12].2, 11[12].8; **ac** RB-B 3[3].4, RB-A 10[11].14, 10[11].16, 11[12].2, 11[12].13, 12[13].4; with poss. pron. 2 sg. **a'th** RB-A LH.7; with poss. pron. 3 sg. m. **a'e** RB-A 12[13].20; with poss. pron. 3 sg. f. **a'e** RB-A <35[36].9>; with obj. pron. 3 sg. m. RB-A 23[24].4; with def. art. **a'r** RB-A 3[3].7, 3[3].9, 10[11].16; 12[13].7, 12[13].24, 24[25].1

a³ preverbal particle: RB-A 1[1].1, RB-B Intro.1, RB-A 2[2].5, RB-B 2[2].1, RB-A 3[3].2; with inf. obj. pron. 3 sg. m. **a'e** RB-B Intro.2, RB-A 23[24].2, 23[24].3; with obj. infixed pron. 3 pl. **a'e** RB-A 52[47].3, Pen.17 96[90].2; with infixed obj. pron. 3 pl. **a's** Add. RBA 248v Bridging Passage.5

a⁴ rel. part. *who, which*: RB-A 2[2].3, 2[2].4, 3[3].14, RB-B 3[3].6, 3[3].17; rel. part. without antecedent RB-A <8[9].2>, 12[13].20; with obj. pron. 3 pl. **a'e** RB-B <35[36].3>, WB <35[36].3>

a⁵ prep. *to*; with def. art. *the*: **a'r** RB-A LH.12, LH.13

a... a... *both... and...*: RB-A <35[36].9>

a dan prep. *under*: RB-A 6[6].4, Pen.17 98[92].5; **y a dan** *from under*: Pen.17 71[66].3–4

aber n.m. *estuary*: RB-A 9[10].10, 23[24].5, RB-B 23[24].3, WB 23[24].4, Ph. 23[24].4

a(c) particle after eq.: RB-A 12[13].1, Pen.17 86[81].7

ac a... rel. pron.: RB-A 60[55].5 (twice), 60[55].6, WB 60[55].6

achaus n.m./f. *cause/event/condition*: Pen.17 74[69].1, **achavs** RB-A 74[69].1; **o achaus** Pen.17 55[50].2, 56[51].3, 92[86].1, 92[86].2; **o achaⱱs** prep. *because, on account of*: RB-A LH.10–11, LH.21

achel n.f. *axis*: Pen.17 94[88].3

acheronta *Acheron*: RB-A 37[37].11

achub pres. ind. 3 sg. *occupy, possess*: RB-A 3[3].11

achubedic verbal adj. *preoccupied*: RB-A LH.10

achwanecka pres. ind. 3 sg. < **achwanecau*** *increase*: RB-A 48[45].1

adaned n. pl. < **adain*** n.f. *wing*: RB-A 10[11].20

adar n. pl. *birds*: RB-A 10[11].9, 10[11].16, 10[11].21, Pen.17 56[51].1, 56[51].2; sg. m. **ederyn** RB-A 12[13].15

adaw* *leave [as gift]*: pres. impers. **edewir** RB-A LH.22; pret. ind. 3 sg. **edewis** RB-A 20[21].3

adaⱱ vn *leave, depart*: Pen.17 74[69].17

adeilo* vn *build, construct, erect*: pret. ind. 3 sg. **adeilwys** RB-A 20[21].1, 20[21].9; **adeilⱱys** RB-A 27[29].3; **adeilvys** RB-A 32[33].6; pret. impers. **adeilut** Pen.17 55[50].2

adeilwyr n. pl. < **adeilwr*** n.m. *builder, founder*: RB-A 30[32].7

adnabot vn *know, recognize*: RB-A 2[2].10, LC.11; pres. ind. 1 sg. **atwen** RB-A LC.6; pres. ind. 3 sg. **atwen** RB-B 32[33].7, WB 32[33].7 RB-A LC.7; pret. ind. 2 sg. **atwaenost** RB-A LH.9

adolỽyn vn *seek, request*: RB-A LC.8
adyrcop *spider*: RB-A <35[36].2>; **adyrcob** WB <35[36].2>; **attyrcob** RB-B <35[36].3>
ae… ae… *either… or…*: RB-A 37[37].15, 48[45].2, Pen.17 71[66].2
Affeliotes *Apheliotes* (wind): Pen.17 60[55].5
Affricus *Africus* (S-W wind): RB-A 60[55].10, RB-B 60[55].10, WB 60[55].7, Rawl. 60[55].10
aflonydu vn *disturb, disrupt*: RB-A LH.11
agorant pres. ind. 3 pl. < **agori*** *open*: Pen.17 57[52].3
allan adv. *from then onwards*: RB-A 11[12].6
Altanus *Altanus* (wind): Pen.17 60[55].15, RB-A 60[55].13, RB-B 60[55].12 (twice), Rawl. 60[55].12 (twice); **Baltanus** WB 60[55].9; **Vltanus** WB 60[55].9
am[1] prep. *around*: RB-A 1[1].3, Pen.17 74[69].10; *upon*: Pen.17 73[68].2; *about* with pron. 3 pl. **amdanunt** Add. RB-A 248v Bridging Passage.7
amaerỽy n.m. *ring, edge, circle*: RB-A 5[5].5, 39[39].2; **amaervy** RB-A 39[39].1
amdiffynn n.m. *defence*: RB-A 12[13].18
amgen adj. *other*: RB-A 2[2].9, 11[12].14; **nyt amgen** grammaticalized phrase *not otherwise, namely*: RB-A 3[3].2, 3[3].11, RB-B 3[3].5, RB-A 29[31].4, 30[32].7; **nyt dim amgen** *not anything other/else*: Pen.17 59[54].2
amgylch n.m. *circle/surrounding area*: Pen.17 93[87].3
amheraỽdyr n.m. *emperor*: RB-A <24[26].2>
amlaf adj. spv. < **aml*** adj. *numerous*: RB-B 23[24].1, WB 23[24].2, Ph. 23[24].1
ampressỽyledic verbal adj. *uninhabited, uninhabitable*: RB-A 6[6].2
amrỽt adj. *raw*: RB-A 10[11].13, 10[11].24
amryuael adj. *diverse, of different kinds*: RB-B 21[22].2, 32[33].7; **amrauaul** RB-A 9[10].14; **amraual** WB 32[33].7; pl. **amryuaelon** RB-A 37[37].14
amser n.m./f. *time*: RB-A 10[11].6, Pen.17 109[103].3, 109[103].4; pl. **amseroed** RB-A 2[2].1, Pen.17 96[90].3
amwregis n.m. *girdle, band*: RB-A 39[39].2
amylder n.m. *multitude, plenty*: RB-A <35[36].7>, LH.20
an prefixed pron. poss. 1 pl. *our*: Pen.17 85[80].2, 85[80].4, RB-A LH.13, LH.22
anadyl n.m./f. *breath, respiration*: Pen.17 71[66].2
anatred see **neidyr**
anednybygedig verbal adj. probably **anednebyddedig*** *unknown, indescribable*: RB-A <35[36].16>
angeu n.m./f. *death*: Pen.17 71[66].3; **agheu** n.m./f. *death*: RB-A 37[37].1
annerch n.m./f. *greeting*: RB-A LC.1
anniueil n.m. *animal*: RB-A 2[2].10 (twice), 12[13].12, 12[13].16; **aneueil** RB-B 2[2].4 (twice); **aniueil** RB-A 12[13].2, 12[13].5, RB-B <35[36].2>; **anniveil** RB-A

<35[36].2>; **anyueil** WB <35[36].1>; pl. **anniueileit** RB-A 9[10].14, 10[11].20, 11[12].1, 11[12].3; **aniueileit** RB-A 3[3].14, 32[33].3, RB-B 32[33].5; **anyueileit** WB 32[33].5; **aniueillyeit** Pen.17 56[51].4, 96[90].4

anniodefedic verbal adj. *unbearable*: RB-A 37[37].13

annỽybot n.m. *ignorance*: RB-A LC.6

annyan n.m./f. *nature, quality, essence*: RB-A 52[47].1, Pen.17 55[50].3, 55[50].5, 56[51].7, 74[69].3; **anyan** RB-A 74[69].3, 74[69].6, WB 74[69].4

annyanaỽl adj. *natural*: RB-A 38[38].1; **anyanaul** Pen.17 73[68].4, 86[81].9

anodun n.m. *depths, abyss*: RB-A 38[38].5, 38[38].6, LH.2

anreithir pres. ind. impers. < **anrheithio*** *plunder/devastate*: Pen.17 74[69].14

anryued adj. *marvellous*: Pen.17 95[89].4

anryỽedaut n.m. *wonder*: Pen.17 128[122].1

ansaỽd n.m./f. *characteristic/nature*: RB-A <8[9].2>

ansodedic verbal adj. *set, located*: RB-A 33[34].1, 74[69].3

anuon vn *send*: RB-A LC.1, LC.9, LH.15; pres. subj. impers. **anỽoner** Pen.17 72[67].5

apiascon *wild parsley* (or other wild umbellifer) < Lat. *apiastrum*: RB-A <35[36].3>

Aquilo (wind): Pen.17 60[55].3; **Aquinilo** RB-B 60[55].2; **Aquil** Rawl. 60[55].2

ar¹ dem. pron. *this one, these*: RB-A 10[11].24

ar² prep. *on, upon, at*: RB-A 1[1].2, 2[2].4, RB-B 3[3].16, RB-A 5[5].1, 10[11].24; with pers. pron. 3 sg. m. **arnaỽ** RB-A 12[13].13; 3 sg. f. **arnei** RB-A 24[25].1, 27[29].4; 1 pl. **arnam** Pen.17 95[89].3; 3 pl. **arnunt** RB-A 11[12].3, 11[12].12, 12[13].11, Add. RBA 248v Bridging passage.7; **arnỽnt** Pen.17 96[90].1; **arnadunt** Pen.17 58[53].4; **arnu** see **y arnu**

ar lun *in the shape of, like*: RB-A 1[1].2

ar warthaf *on top of, on, above*: RB-A 52[47].7, 52[47].8

ar weith *like/in the form of*: RB-B <1[1].2>

arall adj. *another*: RB-B 2[2].4, RB-A 6[6].4, 6[6].8, 7[7].2, 11[12].1; pl. **ereill** RB-A 3[3].11, 5[5].5, <8[9].7>, 9[10].11, 10[11].22

araf adj. *slow*: RB-B 60[55].8, WB 60[55].6, Rawl. 60[55].8

arch n.m./f. *request*: RB-B Intro.2

ardengys pres. ind. 3 sg. < **arddangos*** *display/signify*: RB-A 74[69].17

arderchaỽc adj. *splendid, famous*: RB-A 10[11].4

ardymheredic verbal adj. *temperate*: RB-A 6[6].3, 6[6].7, 60[55].4; **ardymeredic** RB-A 7[7].1, Pen.17 60[55].5, 60[55].9

ardymheru vn *regulate, moderate*: RB-A 5[5].6; *irrigate*: RB-A 52[47].4

ardymherus verbal adj. *temperate*: RB-A 60[55].8

arglỽyd n.m. *lord*: RB-B 2[2].1

arhos vn *wait, remain*: Pen.17 58[53].4

arogleu pl. < **arogl*** n.m./f. *smell, odour*: RB-A 11[12].13, 11[12].15

arthlenn n.f. *body*: RB-A 12[13].5

aruthder n.m. (*cause of*) *fear, wonder*: RB-A 37[37].7

aruthred n.m. *fright, wonder*: RB-A 42[42].3

aruthur adj. *awful, terrible, frightful, strange*: RB-A 12[13].20, 36[37].2, 37[37].15; **aruthyr** RB-A 37[37].12

aruein vn *lead* (in the sense of *bring with*): RB-A 11[12].14

arweddu* *carry, bear, produce*: pres. impers. **arweder** RB-A 29[31].2; pres. ind. 3 sg. **arwed** Pen.17 60[55].12

arỽyd n.m./f. *sign, mark*: RB-A 74[69].18, RB A-516 74[69].4, 74[69].5; **arwyd** RB-A 74[69].18

arwydocaa pres. ind. 3 sg. < **arwyddocau*** *denote, signify*: Pen.17 109[103].4

arỽystyl n.m. *pledge*: RB-A LH.22

aryant n.m. *silver*: RB-A 10[11].7

as obj. pron. 3 pl. RB-A Add. 248v Bridging passage.5

ascỽrn n.m. *bone*: RB-A 12[13].4

assen n.m./f. *donkey*: RB-A 12[13].2

assw adj. *left*: Pen.17 60[55].3, 60[55].6, 60[55].8, 60[55].13, 98[92].4; **asseu** RB-A 60[55].5, 60[55].11, RB-B 60[55].2, 60[55].5, 60[55].7; **assev** WB 60[55].5, 60[55].8

astrologia n.f. *astronomy/astrology*: RB-A 32[33].9

at* prep. *to, towards*: **att** RB-A Add. 248v 88[83].1, 88[83].3, 88[83].4, 88[83].7, 88[83].8; with pers. pron. 1 pl. **attam** RB-A LC.10; **atam** Pen.17 85[80].2; with pers. pron. 3 sg. m. **attaỽ** RB-A 12[13].23, 40[40].2; with pers. pron. 3 pl. **attunt** RB-A 10[11].9

athro n.m. *teacher, master*: RB-A LC.1, LC.2; pl. **athraỽon** Add. RBA 248v Bridging passage.1, Bridging passage.5, RB-A LH.23

atnewydhaer pres. subj. impers. < **adnewyd*** *change/alter in the future*: RB-A 2[2].12

atvo pres. subj. 3 sg. < **atfod*** *to be*: RB A-516 74[69].2

atwneuthur vn *unmake, remake, rebuild*: RB-A <31[32].2>

Aura *Aura* (wind): Pen.17 60[55].15, WB 60[55].9 (twice); **Awra** RB-A 60[55].12; **Aỽra** RB-B 60[55].12 (twice), Rawl. 60[55].11, 60[55].12

Auster *Auster* (wind): Pen.17 60[55].7, WB 60[55].3; **Aỽster** RB-A 60[55].6, 60[55].7, 60[55].8, RB-B 60[55].5, Rawl. 60[55].5; **Awster** RB-B 60[55].7

aỽstralis Latin adj. nominative sg. *South Frigid Zone*: RB-A 6[6].10, 6[6].11

aualeu n. pl. < **afal*** n.m. *apple*: RB-A 11[12].13

awedỽr n.m. *fresh water*: RB-A 49[45].1, 52[47].6

awel n.f. *light wind, breeze*: **awel wynt** set phrase *gust of wind* RB-A 11[12].8

auon n.f. *stream, river*: RB-A <8[9].5>, 9[10].5, 9[10].11, 10[11].1, 11[12].13; **avon** WB 21[22].2, 23[24].1, Ph. 23[24].1, 23[24].3 (twice); pl. **auonoed** RB-A <8[9].5>, 38[38].2 (twice), 48[45].1, 48[45].3; **avonyd** RB-B 38[38].1; **auonyd** WB 38[38].1; **avenyd** Rawl. 38[38].1

aỽr n.f. *hour*: 2[2].11

awen n.m. *muse*: Pen.17 86[81].8

awyr n.m./f. *air*: RB-A 3[3].2, 3[3].3 (twice), 3[3].5, 3[3].8; **aỽyr** RB-A 3[3].5, 3[3].8, RB-B 3[3].14, 32[33].12, Rawl. 38[38].2; **avyr** WB 38[38].2; **aỽyr glan** *pure air, ether* RB-A 1[1].5; **awyr glan** *pure air, ether* RB-A 1[1].6; **aỽyr budur** *unclean air* RB-B <1[1].5>, <1[1].6>; **aỽyr pur** see **pur awyr**

awyraul adj. *aerial*: Pen.17 71[66].1

b

baed n.m. *boar*: RB-A 12[13].6

baril n.m./f. *barrel*: RB-A 74[69].9

barnant pres. ind. 3 pl. < **barnu*** *pronounce judgment (upon a person)*: RB-A 10[11].23

benfygyau vn *borrow (also figurative)*: Pen.17 74[69].4

benfygyedic verbal adj. *lent, borrowed, adopted*: Pen.17 74[69].14

beunyd adv. *continually/always*: Pen.17 74[69].14, RB-A 74[69].13

beis n.m. *ground/bottom*: RB-A 38[38].5

berwi vn *boil, bubble*: RB-A 32[33].5; **berỽi** RB-A 37[37].15

blaen see **ym blaen**

blaenllymhet n.m. *pointedness*: RB-A 12[13].19

bleid n.m. *wolf*: RB-A <35[36].1>, RB-B <35[36].1>; **bleyd** WB <35[36].1>

bleỽ pl. n. *hair, bristles*: RB-A 12[13].9

blodeu pl. n. *flowers*: Pen.17 60[55].12, RB-A 60[55].10, RB-B 60[55].9 (twice), Rawl. 60[55].9; **blodev** WB 60[55].6, 60[55].7

blodeuant pres. ind. 3 pl. < **blodeuo*** *flower, prosper*: RB-A 10[11].7

blỽydyn n.f. *year*: RB-A 10[11].15, 11[12].5, 11[12].6 (twice), 40[40].6; **bluyden** Pen.17 74[69].17; form used with numerals **blyned** WB 74[69].6, Pen.17 81[76].2, RB-A 81[76].3 (twice), 81[76].5; **blynedd** Pen.17 81[76].3

Boreas (wind): Pen.17 60[55].3, RB-A 60[55].2, Rawl. 60[55].2; **Borias** RB-B 60[55].2

bot vn *to be*: RB-B 32[33].8, 32[33].12, WB 32[33].12, RB-A 33[34].1, <35[36].8>; pres. ind. 1 sg. **ỽyf** RB-A LC.8; with prefix *yt-* **ytwyf** RB-A LC.5; pres. ind. 2 sg. with prefix *ytt-* **yttỽyt** RB-A LH.5; pres. ind. 3 sg. **oes** RB-A 5[5].3, 6[6].10, 11[12].13, RB-B 29[31].5, WB 29[31].5; **yv** RB-B 60[55].11, RB-A LC.10; **yw** Pen.17 81[76].1; **yỽ** RB-B Intro.1, RB-A 2[2].12, 3[3].7, 3[3].8, 3[3].9; **yu** Pen.17 55[50].3, 58[53].1, 59[54].1, 60[55].1 (twice); with prefix *ytt-* **yttiỽ** RB-B 74[69].3; **ydiỽ**

WB 74[69].3; **yssyd** RB-A 1[1].2, 1[1].4, 3[3].1, RB-B 3[3].13, RB-A 5[5].2; **ysid** Ph. 21[22].3; **ysyd** Pen.17 59[54].3, 72[67].1, 91[85].1, 95[89].1, 98[92].1; pres. ind. 3 sg. **y mae** RB-A 1[1].2, 1[1].3, 1[1].4, 1[1].5, RB-B <1[1].4>; **e mae** Pen.17 72[67].2, 72[67].3, 73[68].1, 85[80].2, 93[87].3; pres. ind. 3 pl. **y maent** RB-A 10[11].15, 10[11].21, 11[12].7, 12[13].8, 12[13].16; **ynt** RB-A 11[12].8, Pen.17 65[60].1, 96[90].1; consuetudinal pres. ind. 3 sg. **byd** RB-A 1[1].6, RB-B <1[1].3>, RB-A 10[11].5, 11[12].14, 12[13].5, 12[13].15; 3 pl. **bydant** RB-A 11[12].15; rel. form of pres. ind. 3 sg. **yssyd** RB-A 1[1].5, RB-B <1[1].2>, RB-A 7[7].3, RB-B 21[22].3, RB-A 32[33].2; **ysyd** Pen.17 94[88].3; impf. ind. 3 sg. **oed** RB-A 2[2].2, 2[2].3, 10[11].12, 20[21].6, 20[21].11; impf. ind. 3 sg. with prefix *ytt-* **yttoed** RB-A <35[36].9>; pret. 3 sg. **bu** RB-A 2[2].8, RB-B 2[2].1 (twice) 2[2].2, RB-A 20[21].6; pret. 3 pl. **buant** RB-A 34[35].3; consuet. past. ind. 3 sg. **bydei** RB-A 6[6].6, 6[6].7, Pen.17 74[69].8, RB-B 74[69].5, WB 74[69].5; pres. subj. 2 sg. **bych** RB-A LH.4; pres. subj. 3 sg. **bo** RB-A 12[13].7, 38[38].6, 40[40].4, Pen.17 67[62].1, 74[69].10; impf. subj. 3 sg. **bei** RB-A 6[6].3, Pen.17 73[68].4

bratho pres. subj. 3 sg. < **brathu*** *sting*: RB-B <35[36].2>, WB <35[36].2>

breich n.m./f. *arm*: RB-A 3[3].6, 12[13].23; pl. **breicheu** RB-A 12[13].3

brenhin n.m. *king*: RB-A 21[22].1, 21[22].3, RB-B 21[22].1, 21[22].3, RB-A 23[24].4; **brenin** WB 21[22].1, 21[22].3, Ph. 21[22].1, 21[22].3, WB 27[29].3

brenhines n.f. *queen*: RB-A 8[8].2

brenhinyaeth n.f. *kingdom*: RB-A 8[8].3, 20[21].13, 23[24].3, RB-B 23[24].2, RB-A 28[30].2; **brenhinyaet** RB-A 10[11].13; **brenhinaeth** WB 23[24].2; **breniniaeth** Ph. 23[24].2

breuerat n.m. *bleating, bellowing*: RB-A 12[13].20

brwd* adj. *boiling*: **bryttet** RB-B 32[33].3, WB 32[33].3; subst. form **brydyon** RB-B 32[33].8, WB 32[33].8

brumalis Latin adj. nominative sg. *South Temperate Zone*: RB-A 6[6].10

brŵnstan n.m. *brimstone, sulphur*: RB-A 34[35].1, RB-B 34[35].2, RB-A 36[37].3, 37[37].13, 37[37].15; **brwnstan** WB 34[35].1

bryt n.m. *mind*: RB-A LH.10

bwa n.m. *bow, rainbow*: Pen.17 63[58].1; **bua** Pen.17 63[58].3

buanach cpv. < **buan*** adj. *swift, nimble*: RB-A 11[12].8, 12[13].15, 32[33].3

buander n.m. *swiftness, speed*: Pen.17 73[68].3, 95[89].5

buch n.f. *cow*: RB-A LH.13

buched n.m./f. (here f.) *life*: RB-A 2[2].3, <8[9].1>, Pen.17 58[53].2

budugolyaeth n.f. *victory, conquest, triumph*: RB-A 20[21].3

budur see under **awyr**

bŵystuil n.m. *wild beast*: RB-A 12[13].19; **buystuil** Pen.17 140[134].1

buyta *to eat, eating*: Pen.17 71[66].2; impf. ind. 3 pl. **bwytteynt** RB-A 10[11].12; pres. subj. 3 sg. **bŵytao** RB-A <8[9].2>

bych see **bot**

bychan adj. m. *small*: RB-A 1[1].4, 1[1].7

byryant pres. ind. 3 sg. < **bwrw*** *shed*: RB-B 29[31].3

byt n.m. *world*: RB-A 1[1].1, RB-B Intro.1, <1[1].1>, <1[1].5>, RB-A 2[2].2; *sphere*: RB-A 1[1].5; **byt lleihaf** *microcosm*: Pen.17 87[82].4; **byt lleiaf** Pen.17 92[86].6

byth adv. *always, forever*: RB-A <8[9].3>, 29[31].5, RB-B 29[31].3, WB 29[31].3, Pen.17 94[88].2

byu adj. *alive*: Pen.17 55[50].2

<h1 style="text-align:center">c/k</h1>

cablont pres. ind. 3 pl. < **cablu*** *revile/blaspheme*: RB-A LH.13

cadarnhau vn *to reinforce*: RB-A LH.12; pres. ind. impers. **cedernheir** Pen.17 74[69].17

caffel vn *to get*: RB-A <35[36].16>; **caffael** Pen.17 74[69].5; pres. ind. impers. **ceir** RB-A 5[5].7; **keffir** Pen.17 86[81].3; pret. 3 sg. **kauas** RB-B 3[3].14, RB-A 10[11].1, 21[22].1, 21[22].3, RB-B 21[22].1, WB 21[22].1, 21[22].2, Ph. 21[22].1; **kaua[s]** RB-A 8[8].1, 20[21].10; **kafas** RB-B 21[22].2; **kavas** Ph. 21[22].2; **cauas** RB-A 30[32].7; pret. ind. 3 pl. **caѵssant** RB-A 30[32].6; impf. ind. impers. **ceffit** RB-A <35[36].17> (twice); pret. impers. **kaffat** RB-A 32[33].9, 33[34].12, 34[35].3; **caffat** Pen.17 85[80].5, 86[81].3, 86[81].8; pres. subj. impers. **caffer** RB-A 38[38].5

kaer n.f. *stronghold, fortified town/city*: RB-A 20[21].5, RB-B 21[22].3, RB-A 32[33].2, RB-B 32[33].1

calaned pl. < **celain*** n.f. *corpse*: RB-A 10[11].12

Calcias *Calcias* (wind): Pen.17 60[55].6

callaѵr n.m./f. *cauldron*: RB-A 32[33].5; **kallaѵr** RB-B 32[33].8; **kallavr** WB 32[33].8

callon n.f. *heart*: RB-A LH.7, LH.18

canaѵl n.m. *middle*: RB A-516 74[69].5

kan conj. *since, because, before* (< prep. *with*): RB-B 3[3].7; **kann** RB-A 20[21].14, 32[33].9, 37[37].16, LC.5; **can** RB-A LC.7; with copula **ys** *since it is, because it is* **kanys** RB-A 1[1].1, RB-B Intro.1, RB-A 3[3].7, 3[3].10 (twice); **canys** Pen.17 56[51].1, 58[53].2, 60[55].10, 85[80].2, 92[86].2, 94[88].2; **kans** Rawl. 60[55].8; with neg. form of the copula *because it is not* **canyt** RB-A LH.15; with neg. part. **keny** *because ... not*: Pen.17 74[69].9

canhorthѵy n.m. *aid*: RB-A 20[21].3

canmolont pres. ind. 3 pl. < **canmol*** *praise*: RB-A LH.23

cant num. *hundred*: RB-A 81[76].6, Pen.17 88[83].3, 88[83].5, 88[83].7, RB-A Add. 248v 88[83].2; **can** Pen.17 88[83].11, RB-A Add. 248v 88[83].9

cantref n.m. *province*: RB-B 28[30].2, WB 28[30].2

kanys see kan

kanyt see kan

caredig* verbal adj. *diligent*; yn garedic *diligently*: RB-A LH.2

carvan n.m./f. *gum (of teeth), ridge*: RB-A 12[13].4

carỽ n.m. *stag/deer*: RB-A 12[13].3; karỽ 12[13].17; pl. keirỽ 12[13].1

caryat n.m. *love/affection*: RB-A 10[11].22

cestyll pl. < castell* n.m./f. *castle* (here *cloud*): Pen.17 <61[56].2>

cawad* n.f. *shower*: pl. cawadeu Pen.17 <61[56].3>; kaỽadeu RB-A <61[56].3>

caur n.m. *giant/hero*: Pen.17 128[122].1

kedernit n.m. *strength*: RB-A 3[3].12

kedymdeith n.m. *friend*: RB-B Intro.2

kedymdeithas n.f. *bond of friendship*: RB-A LH.22

kedymdeithant pres. ind. 3 pl. *go together*: RB-A 3[3].5

kedymdeithus adj. *companionable*: RB-A LH.6

keinyadaeth n.f. *music/song*: Pen.17 87[82].2

cel* n.m./f. *hiding*, dan gel *in hiding*: RB-A LH.13

kelvydyt n.f. *art, craft*: RB-B 32[33].11; keluydyt WB 32[33].12; pl. keluydodeu RB-A LC.3

kenedyl n.f. *tribe, nation, species*: RB-A 12[13].24, WB 32[33].7; kenedloed n. pl. RB-A 10[11].11, 20[21].13, RB-B 32[33].7

kenllysc plural noun *hail*: Pen.17 60[55].3, RB-A 60[55].2, RB-B 60[55].2, WB 60[55].2, Rawl. 60[55].2, Pen.17 65[60].1, 65[60].2

keny see kan

kerd n.f. *journey*: Add. RBA 248v Bridging passage.1

kerdet vn *walk, go, approach, traverse, move, emanate, flow, extend*: Pen.17 73[68].6, 74[69].16, 74[69].17, RB-A 74[69].14; pres. ind. 3 sg. kerda RB-A 3[3].2, 9[10].4, 9[10].8, 9[10].9 (twice); pres. ind. 3 pl. kerdont RB-A 3[3].14, 11[12].14 (twice), kerdant RB-A 9[10].11, 81[76].5; ipv. 1 pl. kerdun Pen.17 88[83].12; pret. 1 pl. kerdassam RB-A 29[31].8, 32[33].9, <35[36].18>, Pen.17 57[52].3, 71[66].5; kerdo pres. subj. 3 sg. Pen.17 55[50].4, 74[69].14, RB-A 74[69].13, Pen.17 98[92].5

kerwyn n.f. *barrel*: Pen.17 74[69].10

keu adj. *hollow, empty*: Pen.17 63[58].3

keula pres. ind. 3 sg. < ceulo* *congeal*: RB-A 60[55].2

cer prep. *near*: RB-A 74[69].8

cer llaỽ* prep./adv. *near, close to, by*; with 3 pl. ceyr eu llaỽ *close to them, near them*: RB-A <35[36].6>; with 3 sg. m. cer y lav *near it*: RB-A <35[36].12>

kibynn n.m. *shell*: RB-B <1[1].3>, <1[1].4>; kybynn RB-B <1[1].5>

cic n.m. *meat, flesh*: RB-A 10[11].12; kic RB-A 10[11].23, 12[13].15

cilio* *retreat, wane, ebb*: pres. ind. 3 sg. kilia RB-A 40[40].3; kilya Pen.17

74[69].13, WB 74[69].6; pres. subj. 3 sg. **kilio** RB-A 40[40].3, 74[69].12

kilyd n.m. *fellow/companion*: **y gilyd** prep. *together* RB-A 2[2].9, 3[3].3, RB-B 3[3].7, RB-A 12[13].10; **dros y gilyd** *around each other* RB-A 3[3].6; **y gilyd** *the other, another* RB-A 10[11].22

Circius (wind): Pen.17 60[55].2, RB-A 60[55].2

clader pres. subj. impers. < **claddu*** *bury*: RB-A 5[5].7

claf* adj. *sick*: pl. **cleiuon** RB-A <35[36].4>, WB <35[36].5>; **cleifon** RB-B <35[36].4>

claỽr n.m. *surface*: RB-A LC.10

cledyf n.m. *sword*: RB-A 12[13].11

clotuaỽr adj. *famous*: RB-A 20[21].7

clun n.f. *haunch/leg*: RB-A 12[13].2

clust n.m. *ear*: RB-A 12[13].10; pl. **clusteu** RB-A 12[13].4

klybot vn *hear, listen*: Pen.17 85[80].3

kneuit impers. impf. subj. < **cynnau*** *ignite*: RB-A 6[6].4

cnoant pres. ind. 3 pl. < **cnoi*** *chew, torment*: RB-A LH.18 (twice)

knueu n. pl. < **cnu*** n.m. *fleece*: RB-A <35[36].8>

coch adj. *red*: RB-A 12[13].14, 50[46].2 (twice), Pen.17 63[58].5, 74[69].18

cocha pres. ind. 3 sg. < **cochi*** *to redden*: RB-A 74[69].17, RB A-516 74[69].3

cocodrilli n.m. *crocodile*: Pen.17 56[51].5

coet n. pl. *forest*: RB-A 12[13].6

kogyrneu pl. < **cogwrn*** n.m. *shell, husk*: RB-A 12[13].25

colledic verbal adj. *lost*: RB-A <35[36].15>

comedia n.m./f. *comedy*: RB-A 34[35].3

corff n.m. *body*: RB-B 3[3].4, RB-A 5[5].5, 12[13].2, 12[13].13, 12[13].17; pl. **corfforoed** RB-A 10[11].20; **corforoed** Pen.17 58[53].4, 72[67].5

corfforaỽl adj. *corporeal*: Add. RBA 248v Bridging passage.4

korn n.m. *horn, antler*: RB-A 12[13].3, 12[13].18, 12[13].20; *horn of the crescent moon* **korn** RB A-516 74[69].4; pl. *horns* **cyrn** RB-A 12[13].6; **kyrn** RB-A 12[13].10

cornel n.m./f. *corner*: RB-A 74[69].17

Corus *Corus* (wind): Pen.17 60[55].13, RB-A 60[55].11, RB-B 60[55].10, WB 60[55].8, Rawl. 60[55].10

cranc n.m. *crab*: RB-A 12[13].22

creaduryeit pl. < **creadur*** n.m. *creature*: RB-A 2[2].8, Pen.17 58[53].2

credir pres. ind. impers. < **credu*** *believe*: Pen.17 74[69].7, 145[138].2

creedigaeth n.m./f. *creation*: RB-A 2[2].1, 2[2].3, 2[2].5, 2[2].8, 2[2].9

creu vn *create, beget*: Pen.17 56[51].5, 146[139].4; pres. ind. impers. **creir** Pen.17 59[54].1; pret. impers. **crewyt** RB-A 2[2].4, 2[2].6

creula6n adj. *fierce*: RB-A 9[10].14

cribdeillyau *snatch away, remove*: pres. ind. impers. **cribdeilir** Pen.17 73[68].3

cristal n.m. *crystal*: Pen.17 93[87].3

croen n.m. *skin, pelt*: Pen.17 98[92].2; pl. **cr6yn** RB-A 11[12].3

croe6 adj. *sweet, fresh* (of water): RB-A 48[45].2, 52[47].8; cpv. **croewach** RB-A 49[45].2

cr6nn adj. *round*: RB-A 11[12].3, 81[76].1; **krwnn** Pen.17 81[76].1, 96[90].1; f. **cronn** RB-B <1[1].2>, Pen.17 74[69].12

crycha pres. ind. 3 sg. < **crychu*** *to wrinkle*: RB-A 10[11].18

crydu vn *shake*: RB-A <35[36].3>

cryd6st n.f. *tremor, quaking*: RB-A 37[37].7

crymyon adj. pl. < **crwm*** adj. *bent*: RB-A 11[12].3

crynu vn *tremble/quake*: RB-A 42[42].3; pres. ind. 3 pl. **crynnant** WB <61[56].1>

crynua n.m./f. *trembling, gnashing of teeth*: RB-A 37[37].7

cudyau vn *conceal*: Pen.17 95[89].3

cudyedic verbal adj. *hidden*: RB-A LH.3

c6byl adj. *whole/complete*: RB-A LH.5; **kubyl** Pen.17 74[69].12; spv. **c6plaf** RB-A 74[69].11

cumpas n.m. *circle, arc, orbit*: Pen.17 93[87].2

k6n n. pl. < **ci*** n.m. *dog*: RB-A 11[12].4

kwplau vn *finish, fulfil*: RB-A 74[69].16; **k6pla** RB A-516 74[69].2; pres. ind. 3 pl. **kuplaant** Pen.17 81[76].4

k6ynuan n.m./f. *lamentation*: RB-A 37[37].7

cyfarffo pres. subj. 3 sg. *meet*: RB-A 11[12].15

cyuartal adj. *equal*: Pen.17 97[91].2

cyueilornus adj. *wandering*: RB-A LC.7

cyuarthyat n.m. *bark*: RB-A 11[12].4

kyfar6yd adj. *knowledgeable*: RB-A LC.2

kyuaruydyt n.m./f. *guidance/experience/craft*: Pen.17 96[90].1

kyuarwyneb a prepositional phrase *opposite*: RB-A 34[35].4; with def. art.: **kyuarwyneb a'r** RB-A 24[25].2; **kyfarwyneb a'r** RB-B 29[31].1, WB 29[31].1; **kyvar6yneb a'r** RB A-516 74[69].1; *towards, in the direction of* **kyfarwynep a'r** RB-A 23[24].2; **kyfar6yneb a'r** RB-A 74[69].10

kyffelyb adj. *similar, like*: RB-A 12[13].22

kyffelybr6yd n.m. *similarity, likeness*: RB-A 1[1].2; **kyffelypruyd** Pen.17 87[82].3

kyffelybu vn *imitate*: RB-A LH.12

kyffro n.m. *movement, tumult*: RB-A 1[1].1, 42[42].3, Pen.17 73[68].4, 73[68].5, RB-A 74[69].14

kyffro vn *move, excite, stir*: RB-A 42[42].1; pres. ind. 3 sg. **kyffry** Pen.17 55[50].1;

pres. ind. 3 pl. **kyffroant** RB-A 42[42].2

kyffroedic verbal adj. *moved, moving, moveable*: RB-A 1[1].1, Pen.17 57[52].3, 59[54].1

kyflaʋn adj. *complete, full*: RB-A 8[8].4, LH.8

cyuodi *rise, raise*: RB-B 60[55].3, Rawl. 60[55].3; pres. ind. 3 sg. **kyuyt** RB-A <8[9].3>, 9[10].6, 23[24].4, 57[52].2, RB-B 60[55].4; **kyffyt** Rawl. 60[55].4

cyfrʋng prep. *amongst*: RB-B 32[33].4

cyfryʋ adj. *of the same kind, like*: RB-A LH.3; **kyfryʋ** RB-A 8[8].4, 42[42].1; with subordinating conj. in eq. use **kyfryu... ac yd** *of the same kind as*: Pen.17 55[50].2

cyfuch see **uchel**

cyfuerbyn a'r prep. *facing, opposite* + def. art.: RB-A <35[36].5>; **kyuerbyn a'r** RB-A 29[31].1, 33[34].6, Pen.17 74[69].10

cyfun* adj. *agreeing*: **yn gyfun** *in agreement*: RB-A 3[3].7

cyfyrgoll adj. *lost/damned*: RB-A 37[37].5

kyfut n.m. *cubit* (unit of measure): RB-A 10[11].15, 10[11].19, <31[32].3>; **kufyt** RB-A 12[13].19; **kyffut** RB-A <35[36].13>

kyghorvynnus adj. *jealous*: RB-A LH.11; **kyghoruynnus** RB-A LH.17; **kyghorynnʋs** RB-A LH.18

cyhoedd* adj. *public*; **ar gyhoed** *publicly, openly*: RB-A LH.13, LH.14

kyhyt adj. *as long (as), so long (as)*: RB-A 40[40].4

kylch n.m. *circle, rim, orbit*: RB-A 5[5].3, 40[40].6, Pen.17 74[69].2, 74[69].17, RB-A 74[69].2; **kyl[ch]** RB-A 5[5].1; pl. **kylcheu** Pen.17 73[68].1; **kylch y sygneu** *the Zodiac*: Pen.17 86[81].5, 97[91].3; **kylch zodiacus** Pen.17 97[91].2; **ygkylch** prep. *around*: RB-A 1[1].3, 1[1].4 (twice), 1[1].5, 1[1].6 (twice); **yn eu kylch** *around them*: RB-A 3[3].3

kylcha pres. ind. 3 sg. < **cylchu*** *to circle*: RB-B 74[69].6

kyll pres. ind. 3 sg. < **colli** *lose*: RB-A 74[69].12

kymeint adj. *as great, equal*: RB-A 12[13].1, Pen.17 88[83].4; **kemeint** Pen.17 86[81].7; **y gymeint** adj. *as much*: Pen.17 88[83].8

cymell vn *constrain, stimulate*: Pen.17 74[69].15; pres. ind. 3 sg. **kymhellyat** RB-A 74[69].13

cymeryd* vn *take*: pres. ind. 3 sg. **kymer** RB-A 40[40].7, Pen.17 71[66].3; **kemer** Pen.17 72[67].5; pres. ind. 3 pl. **kymerant** RB-A LH.14; **kemerant** Pen.17 58[53].4; pret. ind. 3 sg. **kymerth** RB-A 32[33].8; pres. subj. 3 sg. **kemero** Pen.17 71[66].2

kymherued n.m. *middle/centre*: RB-A 3[3].12, 5[5].2, 5[5].3, 6[6].3, <35[36].13>; **kymerued** Pen.17 97[91].1

kymraec *Welsh* (language): Pen.17 97[91].3

kymydoed n. pl. < **cymwd** n.m. *region/province*: RB-A 38[38].5

cymysc vn *mix, mingle*: RB-A 52[47].6, Pen.17 74[69].7, RB-A 74[69].4, 74[69].7, RB-B 74[69].2; **kymyscu** WB 74[69].3

cymysc adj. *mixed*: Pen.17 74[69].3

kyn... ac eq. construction *so... as*: RB-B 32[33].2–3, 32[33].3, WB 32[33].3, WB 32[33].3–4

kynhebic pres. ind. 3 sg. < **cynhebygu*** *assimilate*: RB-B 2[2].2

kynheliir pres. ind. impers. < **cynnal*** *sustain*: Pen.17 58[53].2

kynn¹ prep. *before*: RB-A 2[2].1

kynn² conj. *although*: RB-A 74[69].9; **kyn** RB-B 74[69].5, WB 74[69].5

cynnal vn *upholding/supporting*: RB-A 5[5].3; **kynnal** RB-B 32[33].12, WB 32[33].12

kynnebonyaỽc adj. *forming an angle in the shape of an elbow* (this is the only example): RB-A 12[13].6

cynt adj./adv. *sooner, previous, formerly*: RB-A 10[11].11, 20[21].14, 34[35].3, <35[36].13>

kyntaf adj. *first, foremost*: RB-B 2[2].1, RB-A 6[6].8, 8[8].2, 21[22].2, 30[32].2; **kentaf** Pen.17 60[55].1, 74[69].2, 92[86].5, 98[92].1

cynullo* vn *to collect, gather*: pres. ind. impers. **kynnullir** RB-A 38[38].1, RB-B 38[38].1; **kynullir** WB 38[38].1, Rawl. 38[38].1; pret. impers. **kynullỽt** Pen.17 92[86].5

kynyruedic adj. *agitated*: Pen.17 59[54].1

kyrchu vn *seek, approach*: RB-A 9[10].10, 23[24].6; **kyrchaf** pres. ind. 1 sg. RB-A LH.19; pres. ind. 3 pl. **kyrchant** RB-A 9[10].13

cyrryeu n. pl. < **cwr*** n.m. *edge*: Pen.17 74[69].19

cysgaut n. *reflection/appearance*: Pen.17 74[69].13; **kysgaỽt** RB-B 74[69].4; **kyskaỽt** WB 74[69].4

kyssondep n.m. *symmetry/harmony*: Pen.17 86[81].8

kystal eq. adj. *comparable/as good* (< **da*** adj. *good/valuable*): RB-A 12[13].8, RB-A <35[36].7>; **kystal a** *comparable to, equivalent to*: RB-A 37[37].10, 37[37].12; **kystal... ac** RB-A 39[39].1

cyt conj. *although*: RB-A 52[47].6, 74[69].13, RB-A Add. 248v Bridging passage.1, Bridging passage.5; **cet** Pen.17 74[69].14

kytredec vn *run together*: Pen.17 95[89].4

kytweda pres. ind. 3 sg. < **cydweddu*** *correspond to*: RB-A 3[3].6, 3[3].8, 3[3].9

kytwledychu vn ? < **cyt*** + **gwledychu*** *feast together* ? : RB-A LC.4

kyỽansodet pret. impers. < **cyfansoddi*** *compose, arrange*: Pen.17 87[82].4

kyỽansodyat n.m. *nature, order*: Pen.17 87[82].2

kyuerbyn a prep. *opposite*: RB-A 33[34].3; with def. art. **kyuerbyn a'r** RB-A 29[31].1, 33[34].6, Pen.17 74[69].10

cyving adj. *narrow/perilous*: RB-A 36[37].3

cyuuna pres. ind. 3 sg. < **cyfuno*** *become one*: RB-A 3[3].9

kywydolyaeth *song, chant*: Pen.17 85[80].1 (twice), 85[80].4, 85[80].5; pl. kywodolaetheu RB-A 12[13].14

kywreinrỽyd n.m. *art, skill*: RB-A 33[34].12

ch

Chimera *Chimera* (constellation): Pen.17 140[134].1

chwech RB-A 28[30].2, 29[31].5, RB-B 29[31].4; chwe RB-A 29[31].5, RB-B 29[31].4; chỽe RB-A 29[31].5; whech num. *six*: RB-B 2[2].3, RB-A 12[13].23, 29[31].6; whe RB-A 2[2].6, 2[2].7; chuechant *six hundred* Pen.17 86[81].6, 88[83].1, 88[83].6; whechant RB-A Add. 248v 88[83].1, 88[83].5

chuechet num. *sixth*: Pen.17 86[81].2

chuedyl n.m./f. *story*: Pen.17 98[92].1

chỽerthin vn *laugh, smile*: RB-A <35[36].3>; wherthin RB-B <35[36].4>; chwerthin WB <35[36].4>

chỽilyaỽ vn *search, seek*: RB-A LH.1

chỽibanat n.f. *whistling, hissing*: RB-A 12[13].14

chwythyat n.m. *breath/act of blowing/hissing*: RB-A 37[37].11; chuythat Pen.17 <61[56].2>; chỽythyat RB-A <61[56].1

chuythat vn *blow*: Pen.17 59[54].2; pres. ind. 3 pl. chuythant Pen.17 60[55].10; chwythant RB-A 60[55].9>

chyỽyl n.m. *proximity*: RB-B 32[33].4, WB 32[33].4

d

da* adj. *good*; yn da adv. *well*: RB-A 2[2].8

daear see **dayar**

dala vn *capture, seize*: RB-A 12[13].11; pres. ind. 3 pl. dalyant RB-B 32[33].5, WB 32[33].4; pres. ind. impers. delhir RB-A 12[13].21

dall adj. here used as substantive *blind*: RB-A LC.6

damwein n.m./f. *accident, event*: RB-A <35[36].16>

damweina vn *come about*: RB-A 12[13].11

damgylchynedic verbal adj. *surrounded/encircled*: RB-A 8[8].6, Pen.17 145[138].2, RB-A LC.8

damgylchyna pres. ind. 3 sg. < **damgylchynu*** *surround*: RB-A 39[39].2

dan gel see **cel***

dan prep. *while*: RB-A <35[36].3>

danned pl. n.m. < **dant*** n.m. *tooth, tusk*: RB-A 12[13].4, 12[13].13, 37[37].7

darffei impf. ind. 3 sg. < **darfod*** *waste away, perish*: RB-A <35[36].16>

darogan vb pres. ind. 3 sg. *prophesy*: Pen.17 74[69].19

da6n n.m./f. *gift*: RB-A LC.3

dayar n.f. *earth, land, Earth* (planet), *Earth* (element): RB-A 1[1].6, RB-B <1[1].6>, RB-A 3[3].2, 3[3].4 (twice); **daear** RB-A 6[6].1, WB 38[38].2, Rawl. 38[38].2, Pen.17 56[51].3 (twice); pl. **dayaroed** RB-A 38[38].3

deall vn *understand*: RB-A Add. 248v Bridging passage.6

dec num. *ten*: RB-A 10[11].5, 29[31].4, 33[34].4, WB 81[76].1, Pen.17 128[122].1; **deg** RB-A 81[76].3, RB-B 81[76].2

dechreu vn *begin, start, originate*: RB-A 30[32].1, Pen.17 146[139].4; impf. subj. 3 sg. **dechreuei** RB-B 81[76].2, WB 81[76].1

dechreu n.m. *beginning*: RB-B 2[2].3, Pen.17 74[69].19, 81[76].3, RB-A 81[76].3

dechymygus pret. ind. 3 sg. < **dychmygu*** *imagine, guess, pretend*: Pen.17 86[81].7

defnyd n.m. *element*: RB-A 1[1].4, 1[1].7, RB-B 2[2].2, RB-A 3[3].1, RB-B 3[3].1; pl. **defnydyeu** RB-A 1[1].3, RB-B <1[1].2>, RB-A 3[3].1, 5[5].4, LH.8; **defnydyon** RB-A LH.19; **defnynnyeu** Pen.17 65[60].1

deggveith < num. **deg*** + n.f. **gwaith*** *ten times*: RB-A 5[5].2

deheu n.m. *south*: RB-A 7[7].4, 9[10].12, 30[32].1, 32[33].1, Pren.17 60[55].2; **dehev** RB-A 10[11].2, 33[34].6, RB-B 60[55].7, WB 60[55].4, 60[55].5

deheuwynt n.m. *south wind*: Pen.17 60[55].7, RB-A 60[55].1, 60[55].4

deil collective n. *leaves, foliage*: RB-A 29[31].5, RB-B 29[31].3, WB 29[31].3

deissyuyt adj. *sudden, abrupt*: Pen.17 71[66].3

deisyueit vn *desire, beg, implore, seek*: RB-A LH.5

deiuyn pres. ind. 3 sg. < **dafnu*** *to drop*: Pen.17 67[62].2, RB-A 67[62].2

delli n.m. *blindness*: RB-A <35[36].4>

del6 n.f. *image*: RB-B Intro.1, RB-A 81[76].4, RB-B 81[76].2, WB 81[76].1, RB-A LH.21; **delw** Pen.17 81[76].3

der6 pret. ind. 3 sg. < **darfod*** *finish*: RB-A 20[21].14, 37[37].16

deng num. *ten*: Pen.17 81[76].2; see also **dec**

dengys pres. ind. 3 sg. < **dangos*** *reveal, signify*: Pen.17 74[69].19, RB A-516 74[69].3

deu num.(m.) *two*: RB-A 3[3].11, 3[3].12, 10[11].5, 10[11].15, 11[12].11

deudec num. *twelve*: RB-A 10[11].19, Pen.17 59[54].2, 81[76].2, RB-A Add. 248v 88[83].2, 88[83].7; **deudeng** Pen.17 81[76].4, RB-A 81[76].2, Pen.17 88[83].2, 88[83].6, 88[83].7; **deudeg** RB-A 81[76].5

deudecet num. *twelfth*: Pen.17 109[103].1

deueit n. pl. < **dafad*** n.f. *sheep*: RB-A <35[36].7>

deugeint num. m. *forty*: RB-A 10[11].13, RB-A Add. 248v 88[83].6, Pen.17 128[122].1

dewissont pres. subj. 3 pl. < **dewis*** *choose*: RB-A 10[11].8

di pron. prefixed 2 sg.: RB-A LH.4

diaereb n.f. *proverb*: RB-A LH.6

dichon* vn *be able, enable, avail*: pres. ind. 3 sg. **dichaỽn** RB-A 12[13].8; **dichaun** Pen.17 85[80].2

didos adj. *sheltered, cosy, snug*: RB-A 12[13].25

diedir pres. ind. impers. < **dyadu*** *to let go, pour*: RB-B 38[38].2

diethyr adj. *outside*: RB-A 1[1].3

dieuyl n. pl. < **diawl*** n.m. *evil/unclean spirits*: Pen.17 58[53].3

diffeith adj. *desert, desolate, uninhabited, barren*: RB-B 29[31].5, WB 29[31].5, RB-A 32[33].4, RB-B 32[33].6, WB 32[33].6

difford adj. *inaccessible*: RB-A 8[8].5; as n.m. *inaccessible place*: RB-A 9[10].14

digriuaf adj. spv. < **digrif*** adj. *pleasant*: Pen.17 85[80].1

diguydau *happen, fall, set* (of sun and moon): Pen.17 92[86].2; pres. ind. 3 sg. **dygỽyd** RB-A 67[62].1, RB A-516 67[62].1; pres. ind. 3 pl. **diguydant** Pen.17 95[89].4

diholet vb pret. impers. < **diol*** *banish*: RB-A 20[21].13

dim n.m. *aught, anything, nothing*: RB-A 5[5].3, 29[31].6 (twice), RB-B 29[31].5, <35[36].15>; **nyt dim amgen** *not anything other/else*: Pen.17 59[54].2

dinas n.m./f. *city, town*: RB-A 10[11].5, 20[21].1, 20[21].9, 20[21].11, 30[32].2

dineu vn *pouring, pour*: Pen.17 143[136].1; pres. ind. impers. **dineuir** RB-B 38[38].2, Rawl. 38[38].1, 38[38].2; **dineir** WB 38[38].1; **deneir** WB 38[38].2; pres. subj. 3 sg. **dineuo** RB-A 81[76].4; impf. subj. 3 sg. **dineuei** Pen.17 81[76].3; **llestri dineu** see **llestri**

diodef vn *allow, endure*: Pen.17 55[50].1

dirgel adj. *secret*: RB-A 52[47].8; **yn dirgel** adv. *secretly* RB-A 9[10].8

dirgeledigyon verbal adj. pl. < **dirgeledig*** verbal adj. *hidden/mysterious*: RB-A 48[45].3

diruaỽr adj. *great, enormous*: RB-A 12[13].9, 12[13].19, 20[21].7, LC.8; **diruaur** Pen.17 73[68].2

disgynnu* *descend, settle*: pres. ind. 3 sg. **diskin** Pen.17 67[62].2; pres. subj. 3 sg. **disgynno** RB-A 37[37].1

distrywu* *destroy, overthrow*: pret. ind. 3 sg. **distrywys** RB-A <31[32].1>; pret. impers. **distrywyut** Pen.17 127[121].2

disynỽyr adj. *senseless*: RB-A LC.12

diwarnaỽt n.m. *day*: RB-A 2[2].6, 2[2].7, RB-B 2[2].3, RB-A 74[69].15; **diwyrnaut** Pen.17 74[69].16

diwed vn *end*: RB-A 81[76].6; **diỽed** WB 81[76].2

dodi* *place, put*: 1 sg. pres. ind. **dodaf** RB-A LH.23; pret. 3 sg. **dodes** RB-A 27[29].4

doeth* n.m. *wise man, sage, the Magi*: pl. **doethon** RB-A <35[36].12>, Pen.17

86[81].8; **doetheon** Pen.17 96[90].4

doethineb n.m./f. *wisdom, discretion*: RB-A LC.1, LC.2, LC.8, LH.1, LH.3

doosparth vn *divide*: Pen.17 97[91].1; pres. ind. impers. **dosperthir** Pen.17 59[54].2, 87[82].1

dosparthedic verbal adj. *divided/separated*: RB-A 1[1].2

dosparthyat n.m. *division*: Pen.17 87[82].1

doui vn *tame*: RB-A 12[13].12, 12[13].21

dreigeu pl. < **draig*** n.m./f. *dragon*: RB-A 37[37].9

dreỽedic n.m. *stinking*: RB-A 37[37].4, LH.13

drewyant n.m. *stench*: RB-A 37[37].13

dros prep. *across*: RB-A 5[5].4; with pers. pron. 3 sg. f. **drosti** RB-A <35[36].10>; **dros y gilyd** *around each other*: RB-A 3[3].6

drud adj. *precious*: RB-A LH.5

dryc adj. m. < **drwg** adj. *bad*: RB-A 11[12].15

dryc ysprytoed *evil spirits*: RB-A 37[37].11

drych n.m. *mirror/form/image*: Pen.17 74[69].4, RB-A 74[69].4, RB-B 74[69].3, WB 74[69].4, RB-A LH.22

du adj. *black*: RB-A 12[13].8, 37[37].10

dua pres. ind. 3 sg. < **duo*** *blacken/darken*: RB-A 10[11].18

duỽ/Duỽ n.m. *God*: RB-A 2[2].2, 2[2].4, 2[2].7, 2[2].8, 5[5].4; **duw** Pen.17 96[90].2

dỽfyr n.m. *water*: RB-A 3[3].2, 3[3].3, 3[3].4 (twice), 3[3].8 (twice); **dỽvyr** RB-A 3[3].3; **dufyr** WB 38[38].1; **duuyr** Pen.17 55[50].3, 56[51].1, 56[51].3, 56[51].4, 56[51].6, 56[51].7; **duỽyr** Pen.17 <61[56].2>; pl. **dyfred** RB-A 5[5].5, Pen.17 56[51].5, 57[52].3, 60[55].4, RB-B <61[56].1>; **dyfyred** WB <61[56].10>; **dỽyfyr** RB-A <61[56].1>

dỽll n.m. *hole*: RB-A 11[12].11

dỽy num. f. *two*: RB-A 6[6].1, 6[6].3, 9[10].11, 10[11].6, 12[13].2; **duy** Pen.17 86[81].6, 88[83].5, 94[88].1

dỽyen dual n. *jaws/cheeks*: RB-A 12[13].6

dỽyn vn *bring/drag*: RB-A LC.7; pres. ind. 3 sg. **dwc** WB 60[55].7; **dỽc** RB-B 60[55].3, 60[55].9, Rawl. 60[55].3; **duc** Pen.17 60[55].12, Rawl. 60[55].9; pret. ind. 3 sg. **duc** RB-A 52[47].3, 98[92].2; pret. impers. **ducpuyt** Pen.17 98[92].2, 140[134].2

dwyrau vn *rise into view/ascend*: Pen.17 92[86].1

dỽyrein n.m. *east*: RB-A 7[7].3, 8[8].3, 9[10].4, 10[11].11, 24[25].2; **duyrein** WB 27[29].2, Pen.17 60[55].11, 73[68].3, 74[69].15, 92[86].1; **dvyrein** WB 32[33].4; **dwyrein** Pen.17 60[55].14, 91[85].1

dỽyreinyaỽl adj. *eastern*: RB-A 23[24].7

dỽyweith adv. *twice*: RB-A 40[40].2

dỽyvron n.f. *breast, chest*: RB-A 11[12].11 (twice); **dỽyuronn** RB-A 12[13].3

dy poss. pron. 2 sg. *your*: RB-A LC.7

dy hun reflexive pron. 2 sg. *yourself*: RB-A LH.9

dyall n.m. *understanding, intellect*: RB-A LH.7

dyall vn *understand*: RB-A Add. 248v Bridging passage.4; pres. ind. 1 sg. dyallͼyf RB-A LH.16; impf. subj. 3 sg. **deallei** RB-A Add. 248v Bridging passage.5

dyd n.f. *day*: RB-A 29[31].6, 29[21].7, RB-B 29[31].4, WB 29[31].3, RB-B 32[33].3; pl. **diheu** RB-B 74[69].5; see also **hanner dyd**

Dyd Braut *Day of Judgment*: Pen.17 58[53].4

dyfraul adj. *watery*: Pen.17 93[87].2, 145[138].3

dygͼyd pres. ind. 3 sg. < **dygwydd*** *flow down*: RB-A 67[62].1, RB A-516 67[62].1

dylyaf pres. ind. 1 sg. < **dylyu*** *should*: RB-A LH.15

dyn n.m./f. *man, human*: RB-A 2[2].10 (twice), RB-B 2[2].3, RB-A 5[5].6, 12[13].5; **den** Pen.17 81[76].4, 86[81].9, 87[82].4; pl. **dynyon** RB-A 10[11].8, 10[11].12, 10[11].15, 11[12].1, 12[13].15; **denyon** Pen.17 58[53].5, 96[90].5, 109[103].3; **deneon** Pen.17 72[67].6, 146[139].1

dyrchauant pres. ind. 3 pl. < **dyrchafu*** *elevate, ascend*: RB-A 11[12].9

dysc vn *learning, teaching*: RB-A Add. 248v Bridging passage.5, RB-A LH.20

dyuot vn *come, issue, emanate*: RB-A <35[36].17>; pres. ind. 3 sg. daͼ RB-A 9[10].3, 10[11].2, 23[24].5, RB-B 27[29].1, WB 27[29].1; **dau** Pen.17 57[52].1; **daw** Pen.17 67[62].1; pret. ind. 3 sg. **doeth** RB-A 27[29].3; **deuth** RB-B 27[29].3, WB 27[29].3; pres. ind. impers. **dewir** RB-A 32[33].9; pres. subj. 3 sg. **del** Pen.17 63[58].1

dyuyat n.m. *growth, form, increase*: RB-A 23[24].2

dyuynder n.f. *depths*: RB-A 37[37].9

dywal adj. *fierce, cruel*: RB-A 10[11].11, 12[13].19

dywedut vn *to say*: Pen.17 81[76].3, RB-B 81[76].3, WB 81[76].2; pres. ind. 3 sg. **dyweit** RB-A 81[76].4, LH.6; pres. ind. impers. **dywedir** RB-A 1[1].1, 2[2].3, 2[2].4, 2[2].7, 2[2].11; **dyvedir** WB 32[33].8; **dyͼedir** WB 74[69].5; impf. ind. impers. **dywedit** RB-A <35[36].17>; pret. ind. 1 sg. **dywedeis** Pen.17 71[66].3

e

ebenus n.m. *ebony*: RB-A <35[36].11>

ebostol n.m./f. *apostle*: RB-A 20[21].12

ebryfygu* vn *forget, despise, neglect*: pres. ind. 3 sg. **ebryuyka** RB-A 37[37].6; pres. ind. 3 pl. **ebryuygant** RB-A LH.12; pret. ind. 3 pl. **ebryuygessant** RB-A 37[37].5

ebryuygedic verbal adj. *forgotten*: RB-A 37[37].5

echtywynnedigrͼyd n.m. *brightness, brilliance*: RB-A 67[62].1, RB A-516 67[62].2

echtywynyca pres. ind. 3 sg. < **echdywynygu*** *be radiant*: Pen.17 72[67].4

edrych vn *to see/regard*: RB-A LH.3; pres. ind. 3 sg. **edrech** Pen.17 96[90].1

edyn n.m./f. *fly*: Pen.17 73[68].5 (twice)

ef affixed pron. 3 sg. m.: RB-A 2[2].3, RB-B 32[33].11, WB 32[33].12, RB-A 50[46].2, RB-A LH.23

ef independent pron. 3 sg. m.: RB-A 5[5].7, RB-B 60[55].1, 60[55].5, 60[55].6, 60[55].9

eglur adj. *clear/visible*: Pen.17 71[66].4, 74[69].20

eglurder n.m. *clarity/brightness/fair weather*: Pen.17 60[55].14, RB-A 60[55].11

ehang adj. *ample*: RB-A 36[37].3

ehedyat vn *flying*: RB-A 12[13].15; pres. ind. 3 sg. **eheta** Pen.17 58[53].3; pres. ind. 3 pl. **ehedant** RB-B 3[3].19, Pen.17 56[51].2

ehelaethach adj. cpv. < **ehelaeth*** adj. *large*: Pen.17 74[69].9

ehun[1] pers. pron. 3 sg. **e** + **hun**: m. *himself*: RB-A 6[6].11, Pen.17 96[90].2, RB-A LH.14, *its own* RB-B 2[2].4, Pen.17 55[50].5; f. *herself*: Pen.17 74[69].4

ehun[2] pers. pron. 3 pl. + reflexive: *themselves*: RB-A 48[45].3; **ehunein** RB-A LH.17, LH.18

eigaⱱn n.m./f. *ocean*: RB-A 38[38].5, 49[45].2; **eigyaⱱn** RB-A 39[39].1, 50[46].1

eil num./adj. *second*: RB-A 2[2].3, RB-B 2[2].1, RB-A 6[6].9, 9[10].5, 38[38].1

eilwers adv. *afresh*: **pob eilwers** see **pob**

eiry n.m. *snow*: Pen.17 60[55].3, RB-A 60[55].2, RB-B 60[55].2, WB 60[55].2; ; **eira** Rawl. 60[55].2

eisseu n.m./adj. *lack, less* (used in composite numbers): RB-A 40[40].6, 74[69].16; **eissyeu** Pen.17 74[69].17

eissoes adv. *nevertheless, however*: RB-A 50[46].1, 52[47].1, 74[69].14, RB-A Add. 248v Bridging passage.6; **eissyoes** Pen.17 73[68].3, 73[68].5, 74[69].15, 95[89].1

eistedua n.f. *abode/throne*: RB-A Add. 248v, Bridging passage.3 (twice)

eithaf adj. *extreme, outer, most distant*: RB-A 6[6].1, 32[33].5, RB-B 32[33].8, WB 32[33].9, RB-A 36[37].2

either prep. *beside*: Pen.17 60[55].14; **eithyr** RB-A 60[55]12

eliphant n.m. *elephant*: RB-A 12[13].6, 12[13].17; pl. **eliphanyeit** RB-A 12[13].24

emrithyassant pret. 3 pl. < **emritho** *shape, form oneself*: Pen.17 109[103].2

en see **yn**

ena see **yna**

eneit n.m./f. *soul, spirit*: Pen.17 87[82].3; pl. **eneideu** RB-A 37[37].3; **eneidyeu** Pen.17 146[139].3

engylyon n. pl. *angels*: Pen.17 72[67].5; **egylyon** RB-A Add. 248v Bridging passage.3; **Brenhyn er Englylyon** *King of the Angels* (= God) Pen.17 147[140].1–2

ennwir *wicked*: RB-A 10[11].23

eno see **yno**

enteu[1] adv. *therefore, then*: Pen.17 56[51].2, 56[51].3, 56[51].4, 65[60].1, 67[62].2

enteu[2] see **ynteu**

enỽ n.m. *name*: RB-A 8[8].1, 10[11].1, 20[21].9, 20[21].10, 21[22].1; **enw** RB-A 21[22].3, WB 21[22].1, Pen.17 59[54].2, 92[86].5, 96[90].3; pl. **enỽeu** RB-A 30[32].6; **enweu** Pen.17 96[90].4

enỽi vn. *call, name*: RB-B 2[2].5; pres. impers. **ennwir** RB-A 10[11].3; **ennỽir** RB-A 10[11].10, 33[34].2; **enwir** Pen.17 92[86].3, 93[87].1; pret. 3 sg. **enwis** Pen.17 96[90].4; pret. impers. **ennwit** RB-A 20[21].9; **ennỽit** RB-A 20[21].2, 20[21].4, 20[21].7; **enỽit** WB 27[29].2; **enwit** RB-B 27[29].2, 32[33].9, 32[33].10, WB 32[33].10

equinoctialis Lat. *Torrid Zone*: RB-A 6[6].9

erbyn prep. *opposite, in front*: RB-A 12[13].20; see also **yn erbyn**

ereill see **arall**

escop n.m. *bishop*: RB-A <31[32].4>

eskyllwynnyeu n. pl. < **asgellwynt** n.m. *side-wind, lesser/minor wind*: Pen.17 59[54].4

esgynnu *rise, ascend*: pres. ipv. 1 pl. **esgynnỽn** Pen.17 57[52].4; **esgynnwn** Pen.17 71[66].5; pret. 1 pl. **esgynnassam** Pen.17 88[83].11

ettwa adv. *again*: RB-A 2[2].12

eu poss. pron. 3 pl. *their*: RB-A 2[2].6, 3[3].3, 3[3].5, 6[6].1, 9[10].12

eur n.m. *gold*: RB-A 10[11].7, 10[11].8, 20[21].7, Pen.17 74[69].18, RB-A 74[69].17

eureit adj. *golden*: RB-A <35[36].8>

eurin adj. *golden/splendid*: Pen.17 98[92].2

Euroauster *Euroauster* (wind): Pen.17 60[55].8; **Euro Auster** WB 60[55].4; **Eỽro** RB-A 60[55].7; **Euro** RB-A 60[55].7; **Eurus** RB-B 60[55].2, 60[55].5, Rawl. 60[55].5

Euronothus *Euronothus* (wind): Pen.17 60[55].9, WB 60[55].5; **Euromothus** RB-B 60[55].7; **Eure Nothus** Rawl. 60[55].7

euyd n.m. *bronze, copper*: Pen.17 81[76].3, RB-A 81[76].4; **evyd** RB-B 81[76].2; **efuyd** WB 81[76].1

ewined pl. < **ewin*** n.m./f. *talon/claw*: RB-A 10[11].20, 11[12].3

f/ff

Fauonius *Favonius* (wind): Pen.17 60[55].11

fflamaỽl adj. *fiery*: RB-A LC.9

ffo vn *flee, escape*: Pen.17 109[103].1; pres. ind. 3 sg. **ffo** RB-A 10[11].18; pres. ind. 3 pl. **ffoant** RB-A <35[36].5>

fford n.f. *road, way*: RB-A 5[5].5, 11[12].14, 52[47].7; pl. **ffyrd** RB-A 48[45].3

ffurfỽyt pret. impers. < **ffurfio*** *to form/create*: RB-B 2[2].2

ffuruauent n.f. *firmament, heavens*: RB-A 1[1].5; **furuauent** Pen.17 72[67].2, 85[80].4; **ffurua6ent** Pen.17 74[69].15; **ffur6auent** Pen.17 73[68].3; **ffur6auen** Pen.17 88[83].8, 95[89].3; **ffuruauen** RB-A 74[69].14, WB 74[69].6, Pen.17 86[81].4, RB-A 248v Bridging passage.2, 88[83].8; **ffuruaven** RB-B 74[69].6

ffuruf n.m./f. *form, shape*: RB-A 1[1].2, LH.7; **ffuryf** RB-A 2[2].2; **furyf** RB-A 2[2].9, Pen.17 93[87].2

ffuryfedigaeth n.f. *formation*: RB-B <1[1].1>

furyfaedic verbal adj. *formed*: Pen.17 93[87].1

ffr6yth n.m./f. *fruit, produce*: RB-A <35[36].15>

ffr6ythla6n adj. *fertile*: RB-A 10[11].7

ffrydyeu n. pl. < **ffrwd*** n.m./f. *stream*: RB-A LC.2

ffyd n.f. *faith*: RB-A LC.2

ffynna6n n.f. *fountain, spring*: RB-A <8[9].3>, 9[10].11, 11[12].12, RB-B 32[33].2, RB-A 52[47].5 (twice); **ffynnawn** WB 32[33].3; pl. **ffynhonneu** RB-A <35[36].4>; **fynhonnev** RB-B <35[36].4>; **ffynnhonev** WB <35[36].4>; **ffynnhonneu** RB-A 38[38].2; **fynhonnyeu** Pen.17 55[50].2

g

gallu n.m. *ability*: Pen.17 85[80].3

gallu vn *to be able to*: RB-A Add. 248v Bridging passage.6; pres. ind. 3 sg. **geill** RB-B 32[33].3, WB 32[33].3, 32[33].4; pres. ind. 1 pl. **gallwn** Pen.17 85[80].3, 94[88].2; pres. ind. 3 pl. **gallant** Pen.17 56[51].4; pres. ind. impers. **gellir** RB-A 6[6].1, 6[6].5, 10[11].9, 12[13].12, 12[13].21 (twice); impf. ind. 3 pl. **gellynt** RB-A Add. 248v Bridging passage.5; impf. subj. impers. **gellit** RB-A 2[2].10, RB-A Add. 248v Bridging passage.7; pres. subj. 2 sg. **gellych** RB-A LH.7; pres. subj. 3 sg. **gallo** Pen.17 96[90].5; pres. subj. 3 pl. **gallont** RB-A LH.11, LH.12; impf. subj. 3 sg. **gallei** RB-B 81[76].3, WB 81[76].2

galw* *call* (*by name*): pres. ind. impers. **gelwir** RB-B Intro.1, RB-A 7[7].4, 9[10].2, 9[10].6, 9[10].13; **gelwyr** RB-A 37[37].10; **gel6ir** WB 32[33].5, RB-A 34[35].5, Rawl. 60[55].2, 60[55].3, WB 81[76].3; pret. 3 sg. **gelwis** RB-A 20[21].2; pret. impers. **gelwit** RB-B 3[3].1, RB-A 20[21].5, 20[21].9, 23[24].4, <31[32].2>; **gel6it** RB-A 20[21].5; **gelwyt** RB-A 27[29].2

gama Lat. *musical scales*: Pen.17 86[81].1

gan prep. *with, on account of*: RB-A 10[11].18, 48[45].2 (twice), Pen.17 74[69].16, 85[80].1, RB-A LH.17; with pron. 3 sg. m. **ganta6** RB-A 12[13].5, 12[13].20; **y gan** prep. (< prep. **y** *from* + **gan** *from*) *on account of, from*: RB-B 2[2].3 , 2[2].4 (twice), RB-A 8[8].1, 10[11].1; **e gan** Pen.17 74[69].4, 74[69].14; with pers. pron. 3 sg. f. **y genti** RB-B 32[33].9; **y genthi** WB 32[33].10; **genti** RB-B 74[69].3; with pers. pron. 1 pl. **y gennym** Pen.17 94[88].2

gauyr n.f. *goat*: Pen.17 140[134].2, RB-A LH.6

gayaf n.m. *winter*: RB-A 6[6].4, 29[31].5, RB-B 60[55].9, WB 60[55].7; **gaeaf** RB-A 29[31].6, Pen.17 60[55].12, Rawl. 60[55].9, 98[92].4

geneu n.m. *mouth, lips, pass*: RB-A 11[12].11, 12[13].4, 12[13].10, 36[37].3, 37[37].10; **genev** RB-B <35[36].3>, WB <35[36].3>

geni vn *to be born, birth*: RB-A LH.16; pres. ind. impers. **genir** RB-A 11[12].5, <35[36].1>

glan adj. *pure*: RB-A 1[1].5, 1[1].6

glas adj. *blue, greenish blue*: Pen.17 63[58].5; pl. **gleisson** RB-A 12[13].14

glaỽ n.m. *rain, shower*: RB-A 49[45].1, 67[62].3, RB A-516 67[62].2; **glaw** Pen.17 65[60].1, 67[62].3

glawauc adj. *rainy, wet*: Pen.17 74[69].20, 109[103].5; **glaỽaỽc** RB-A 74[69].18, RB A-516 74[69].4

gloewach adj. cpv. *as bright* < **gloyw*** adj. *bright*: Pen.17 72[67].2, 72[67].3, 72[67].4

gloyỽ n.m. *clearness* (here as synonym for *ether*): RB-B <1[1].6>

gogled n.m. *north*: RB-A 7[7].4, 21[22].2, 23[24].3, 24[25].1, 27[29].1

gogledwynt n.m. *north wind*: Pen.17 60[55].1, RB-A 60[55].2

goleu n.m. *light, clearness*: Pen.17 74[69].12, RB-A 74[69].10, RB A-516 74[69].1

goleuat n.m. *light*: RB-B 60[55].11, WB 60[55].8, Rawl. 60[55].11, Pen.17 72[67].5, 74[69].6

goleuhau* vn *shine, glitter, give light*: pres. ind. 3 sg. **goleuhaa** Pen.17 74[69].11, RB-B 74[69].3; **golevhaa** WB 74[69].4; impf. subj. 3 sg. **goleuhaei** Pen.17 74[69].7, RB-A 74[69].7; pres. subj. 3 sg. **goleuhet** RB-B 74[69].5; **goleuet** WB 74[69].5

golỽc n.m./f. *eyesight*: RB-A LH.7

gordyuynder n.m. *profundity, great depth*: RB-A LH.2

gorffenno pres. subj. 3 sg. < **gorffen*** *end*: RB-A LC.3

gorffỽwys* *rest, allow to rest*: pret. 3 pl. **gorffỽyssant** RB-A LH.11; pres. subj. 3 pl. **gorffỽyssont** RB-A 11[12].9

gorfod* *be victorious/prevail*: pres. ind. 3 sg. **gorỽyd** Pen.17 95[89].2; **gorvyd** RB-A LH.19

gorllewin n.m. *west*: RB-A 7[7].3, 7[7].4, 23[24].2, 28[30].1, 29[31].1; **gorlleỽin** RB-A 10[11].3, 27[29].4; **gvrllewin** WB 29[31].1; **gorllevin** WB 32[33].9; **gollewin** Pen.17 60[55].5, 73[68].3, 74[69].15, 91[85].1, 92[86].1

gormod n.m. *excess*: Pen.17 71[66].1

gorthrum n.m. *oppression/heaviness*: Pen.17 67[62].1

gorthrymu* vn *oppress/grow heavy*: pres. subj. impers. **gorthrymer** RB-A 67[62].1; **gỽrthrymher** RB A-516 67[62].1

goruchelyon adj. pl. < **goruchel*** *highest*: RB-A LH.19

gorwed pres. ind. 3 sg. < **gorwedd*** *lie, be buried*: Pen.17 98[92].3

gosged n.m./f. *aspect/shape*: RB-A 12[13].12

gossodedic verbal adj. *situated*: RB-A Add. 248v Bridging passage.2, Pen.17 95[89].1

gossodet pret. impers. < gosod* *establish*: Pen.17 98[92].3

graỽn pl. n. *grain, seed*: RB-A 10[11].18

griffyt n.m. *griffon*: RB-A 10[11].9, 10[11].16; pl. griffonnyeit RB-A 10[11].19

groec *Greek language*: Pen.17 92[86].4, 97[91].3

grymhau vn *become strong*: RB-A LH.2

gwac adj. *empty*: Pen.17 58[53].1

gỽadyn n.m./f. *heel, sole of foot*: RB-A 11[12].9; pl. gỽadneu RB-A 11[12].2

gwaelaỽt n.m. *bottom, depths*: RB-A <35[36].13>, 36[37].3, 38[38].6; gỽaelaỽt RB-A 49[45].2, 74[69].9; gỽaelaỽvt 52[47].9; guaelaut Pen.17 74[69].10

guaet n.m. *blood*: Pen.17 55[50].4

gỽaeỽ n.m./f. *spear, lance*: RB-A 12[13].11

gwahanu* *separate, section*: pres. ind. 3 sg. gỽahana RB-A 9[10].10; gwahana RB-A 38[38].4; gwahan RB-B 38[38].3, WB 38[38].3; pres. ind. 3 pl. gỽahanant RB-A 23[24].6; pres. ind. impers. gwehenir RB-A <8[9].4>; gỽehenir RB-B 38[38].2; gvehenir WB 38[38].2; pret. 3 sg. gwahanỽys RB-A 2[2].8; gỽahanỽys RB-A 52[47].3; pret. impers. gwahanỽyt RB-B 2[2].3; gỽahanỽyt RB-A LH.8

gwahannedic adj. *separate*: RB-B <1[1].2>

gwahanredaỽl adj. *separate/various*: RB-A 2[2].6, RB-A Add. 248v Bridging Passage.3; guahanredaul Pen.17 73[68].1

gwarandau vn *listen to, hear*: Pen.17 85[80].3

gwarchae vn *besiege/surround*: RB-A 10[11].11

gỽarchaedic verbal adj. *surrounded/confined*: RB-A 42[42].2

gwascaỽt n.m. *shade, umbrella*: RB-A 11[12].9; gỽascaỽt RB-A 74[69].12

gwasgaru* *scatter, separate, rout, disperse, waste*: impers. pres. gwasgerir RB-A 38[38].2; gỽesgerir RB-A 38[38].3

gwdant see gwybot²

gweddu* vn *be yoked together*: pres. ind. 3 sg. gỽeda RB-B 3[3].8, 3[3].11; gweda RB-B 3[3].6, 3[3].10; *befit, serve (the purpose)* pres. ind. 3 pl. gwedant Pen.17 86[81].9

guedwheir pres. ind. impers. < gweddwhau* *deprive*: Pen.17 74[69].13

gỽedy¹ conj. *after*: RB-A 10[11].23, RB-A <31[32].2>, LC.3; gwedy Pen.17 97[91].1, RB-A LC.11; guedy Pen.17 146[139].2; gwedy na *after... not...* Pen.17 109[103].2

gỽedy² prep. *after*: RB-B 3[3].5, RB-A 20[21].3, Ph. 23[24].5, RB-B 81[76].2, WB 81[76].1; gwedy WB 23[24].5, Pen.17 65[60].1, RB-A 81[76].3, 81[76].5

gỽefleu n. pl. < gwefl* n.f. *lip, jaw*: RB-A <35[36].3>

gweilgi n.m. *flood, sea, the deep*: RB-A 12[13].24, 52[47].9, Pen.17 60[55].15, RB-A 60[55].12, RB-B 60[55].12

gweithret n.m./f. *action, work*: RB-A LH.14; gueithret Pen.17 55[50].4; pl. gweithredoed RB-A 2[2].7

gwelet vn *see*: RB-A LH.7; pres. ind. 1 pl. gỽelỽn RB-A LC.11; pres. impers. gỽelir

WB 74[69].2, RB-A LC.6; **gwelir** Pen.17 58[53].1, 74[69].6, RB-A 74[69].5, WB 74[69].4, Pen.17 94[88].1; **guelir** Pen.17 74[69].1; **gwelhir** RB-B 74[69].4; pret. impers. **gỽelet** RB-A 74[69].1, 74[69].9, LH.21; **guelet** Pen.17 94[88].2; pres. subj. impers. **gweler** Pen.17 57[52].2, 74[69].9; impf. subj. 3 sg. **gwelei** WB 81[76].1

gwell adj. *better*: RB-A LH.9

gỽellt n.m. *grass*: WB 60[55].6

gỽennỽynic adj. *poisonous, venomous*: RB-A 29[31].3

gwenwyn n.m. *poison, venom*: Pen.17 71[66].3

gwerthyt n.f. *axis, shaft*: Pen.17 94[88].1

guibyauder adj. *wandering*: Pen.17 73[68].2

gỽlan n.m. *wool*: RB-A <35[36].7>

gỽlanha vn *gather wool*: RB-A LH.6

gỽlat n.f. *land, region, kingdom*: RB-A 9[10].9, 10[11].1, RB-B 21[22].1, WB 29[31].2, RB-A 30[32].2; **gwlat** RB-A 20[21].3, WB 21[22].1, RB-A 29[31].2, RB-B 29[31].2, 29[31].3; **gỽlad** Ph. 21[22].1; pl. **gwladoed** RB-B 38[38].3; **gỽladoed** WB 38[38].3

gỽeryt n.m. *earth, soil*: RB-A 29[31].2

gwisc vn *clothes*: RB-A 11[12].3

gỽled n.f. *repast*: RB-A 10[11].23

gulith n.m. *dew*: Pen.17 67[62].1, 67[62].3; **gỽlith** RB-A 60[55].10, RB-B 60[55].9, Rawl. 60[55].9, RB-A 67[62].1, 67[62].3 (twice)

gỽlyb adj. *wet*: RB-A 3[3].8 (thrice), RB-B 3[3].9; **gwlyp** Pen.17 56[51].3; **gulyp** Pen.17 58[53].2

gỽlyb n.m. *wetness*: RB-B 3[3].10

gulybur n.m. *liquid, moisture, dampness*: Pen.17 60[55].7; **gỽlybỽr** RB-A 60[55].6; **gulybỽr** RB-B 60[55].6; **gỽlybvr** Rawl. 60[55].6; **gvlybỽr** WB 60[55].4

gỽneuthur vn *do, accomplish, fashion, make*: RB-A 10[11].23, LC.11; **guneuthur** Pen.17 56[51].2; pres. ind. 1 sg. **gỽnaf** RB-A 2[2].13; pres. ind. 3 sg. **gỽna** RB-A 6[6].8, 50[46].2, RB-B 60[55].10, WB 60[55].1, 60[55].2, 60[55].3; **gwna** RB-A 10[11].5, Pen.17 60[55].2, 60[55].4, RB-A 60[55].1, 60[55].2; **gvna** Rawl. 60[55].10; pres. ind. 3 pl. **gwnant** Pen.17 60[55].3, 60[55].9; pres. impers. **gỽneir** RB-A 12[13].25, LH.28; impf. ind. 3 sg. **gwnaei** Pen.17 63[58].4; pret. ind. 3 sg. **gwnaeth** RB-A 2[2].5, 20[21].11, RB-B 32[33].9, <35[36].12>; **gỽnaeth** RB-B Intro.2, RB-A 2[2].7, RB-B 32[33].11; **gvnnaeth** WB 32[33].11; **gvnaeth** WB 32[33].12; **gwnnaeth** WB 32[33].9; pret. impers. **gwnaethpỽyt** RB-A 2[2].3, RB-B 3[3].3; **gỽnaethpỽyt** RB-A 3[3].1, RB-B 3[3].2, RB-B 3[3].17; **gỽnaetpỽyt** RB-B 3[3].4; pres. subj. 3 sg. **gwnel** RB-A 10[11].24; pres. subj. impers. **gwneler** Pen.17 55[50].4; impf. subj. 3 sg. **gwnelei** RB-B 81[76].2

gỽr n.m. *man*: RB-B 27[29].3, WB 27[29].3; pl. **gỽyr** RB-A <31[32].2>

gvregis n.m. *belt, zone, circle*: RB-A 39[39].1; **gỽregys** RB-A 40[40].1; pl.

gỽregysseu RB-A 39[39].1

gỽreic n.f. *woman*: RB-A 32[33].7

gureigyaul adj. *pertaining to a woman*: Pen.17 55[50].4

gỽres n.m. *heat*: RB-A 6[6].3, 6[6].8, RB-B 32[33].6, 60[55].6, WB 60[55].4, Rawl. 60[55].6; **gwres** RB-A 6[6].5, 6[6].6, 32[33].5, RB-B 32[33].8, WB 32[33].8, RB-A 37[37].13; **gvres** WB 32[33].6; **gures** Pen.17 71[66].1

guressaei impf. ind. 3 sg. < **gwresau*** *to heat, warm*: Pen.17 74[69].8; **gỽressaei** RB-A 74[69].8

gỽressaỽc adj. *hot*: RB-A 3[3].9, 60[55].7; **gỽressavc** RB-B 3[3].10; **guressauc** Pen.17 60[55].8

gwrthỽyneb adj. *contrary, opposite*: **gỽrthỽyneb** RB-A 12[13].9; **gurthuynep** Pen.17 55[50].3; **yg wrthỽyneb** adv. *conversely*: RB-A 3[3].4; **yg gurthuynep** Pen.17 74[69].15, 81[76].1; **y urthuynep** Pen.17 63[58].4

gwrthwynebu* *resist/oppose*: pres. subj. 3 sg. **gỽrthỽynepo** RB-A 12[13].20; **gurthuynepo** Pen.17 63[58].2

gỽrych n.m. *bristles*: RB-A 11[12].12

gỽrychonen n. sg. < **gwrychion*** n. pl. *sparks*: RB-A LC.9

gwybot[1] n.m. *knowledge, wisdom*: RB-A LC.1, LC.9; **gỽybot** RB-A LH.2

gwybot[2] vn *knowing*: RB-A LC.6; pres. ind. 3 sg. **guyr** Pen.17 96[90].2; **gwyr** Pen.17 96[90].3; pres. ind. 3 pl. **gwdant** RB-A LC.6

gwydd* n.m. *face*; **wyd y gwyd** *face to face*: RB-A LH.4

gwyd collective n. *twigs, forest*: RB-A 29[31].4, RB-B 29[31].3; **gỽyd** WB 29[31].3

gỽyllt adj. *wild* (of animals): RB-B 32[33].5, <35[36].2>; **gvyllt** WB 32[33].6, <35[36].2>; **gỽyllt** *stormy* (of weather): RB-A 40[40].5; cpv. **gỽylltach** RB-A 40[40].4

gwyn n.m. *white* (of an egg): RB-A 1[1].4; **gỽynn** RB-A 1[1].3, RB-B <1[1].3> (twice), <1[1].5>, <1[1].6>

guynn adj. *white*: Pen.17 43[136].1; spv. **gỽynaf** RB-A 33[34].10; pl. **gỽynnyon** RB-A <35[36].7>; **yn wynn** predicative *in the state of being white*: RB-A 10[11].17

gỽynha pres. ind. 3 sg. < **gwynhau*** vn *whiten*: RB-A 67[62].3, RB A-516 67[62].2

gwynnheu n.m./f. *pain, ache*: RB-A 37[37].15

gỽynt n.m. *wind*: RB-A 20[21].8, Pen.17 60[55].5, RB-A 60[55].5, 60[55].8, RB-B 60[55].7; **guynt** Pen.17 57[52].2, 59[54].1, 60[55].11, 60[55].12; **gwynt** Pen.17 60[55].15, RB-A 60[55].3, WB 60[55].9, RB A-516 74[69].3; **awel wynt** set phrase *gust of wind*: RB-A 11[12].8; pl. **gwynneu** RB-A 42[42].1; **gỽynneu** RB-A 42[42].2, 60[55].8; **gỽynne** RB-A <61[56].1>; **guynnyeu** Pen.17 55[50].1, 59[54].1, 60[55].9; **gỽynnoed** RB-B <61[56].1>, WB <61[56].1>

gỽynuydedigrỽyd n.m. *beatitude, blessing*: RB-A LC.4

guyrd adj. *green, grey*: Pen.17 63[58].5

gỽyrtheu n. pl. < **gwyrth*** n.m./f. *miracle*: RB-A 5[5].4

gwyth* n.f. *vein/stream*: n. pl. **gỽythi** RB-A 5[5].6; **gwythi** RB-A 48[45].2

gwywyant pres. ind. 3 sg. < **gwywo*** *wither*: WB 29[31].3

gyr llaỽ prep. *near, close to*: RB-A 9[10].6–7, 9[10].10; **geyr llaỽ** RB-A 11[12].12, 34[35].3–4; **ger llaỽ** RB-A 27[29].3

gyrdder n.m. *fierceness, cold*: Pen.17 67[62].1

gyrru* *to compel, drive off*: pres. ind. 3 sg. **gyrr** RB-B 60[55].9, WB 60[55].7; **gyr** Rawl. 60[55].9; **gyrrant** pres. ind. 3 pl. RB-B <61[56].1>

(y) gyt (a) prep. *together with*: **gyt a'r** RB-A LC.5; **gyt ac** prep. *together with*: RB-A 27[29].3; **y gyt** adv. *together*: RB-A 2[2].5; **y gyt a(c)** prep. *together with*: RB-B 27[29].3, WB 27[29].3, RB-A LC.8; **y gyt a'r** prep. *together with the*: Pen.17 146[139].4, RB-A LC.6, LC.12

h

haf n.m. *summer*: RB-A 10[11].5, 29[31].5, 29[31].6, Pen.17 98[92].4

hagen conj./adv. *however, moreover*: RB-A 38[38].6, Pen.17 74[69].1, 74[69].7, RB-A 74[69].6, Pen.17 96[90].4

hallt adj. *salt, salty*: RB-A 48[45].2, 49[45].1, 52[47].1 (twice), 52[47].8

hanfod* *be from*: pres. ind. 3 sg. **henyỽ** RB-A 52[47].4; pret. 3 sg. **hanoed** RB-A 33[34].11; pres. subj. 3 sg. **hannffo** RB-A 52[47].2

hanner n.m. *half*: Pen.17 86[81].5 (twice), 86[81].7 (twice), RB-A Add. 248v 88[83].3

hanner dyd *midday*: Pen.17 92[86].32; *south* RB-B 32[33].1, WB 32[33].1; see also **meridies**

haỽd adj. *easy*: RB-A Add. 248v Bridging passage.4

heb prep. *without*: RB-A 11[12].10, 29[31].6 (twice), LC.11; **hep** Pen.17 71[66].4

helaethach cpv. adj. *larger* < **helaeth*** adj. *large*: RB-A 74[69].8

heli n.m. *brine*: RB-A 10[11].25

henhau* *grow old, age*: pres. ind. 3 pl. **henhaant** RB-A 10[11].16; pres. subj. 3 pl. **henhaont** RB-A 10[11].23

hergorn vn *transfix*: RB-A 12[13].20

herỽyd prep./conj. *according to, because*: RB-A 2[2].6; **herwyd** Pen.17 72[67].2, 98[92].1

heul n.m./f. *sun, Sun (as planet)*: RB-A 6[6].8, 29[31].3, 32[33].4, 32[33].5, RB-B 32[33].6

heuyt adv. *also, in addition, likewise*: RB-A 9[10].2, 10[11].13, 10[11].21, 11[12].4, 12[13].12

hi affixed pron. sg. f. *she, her*: RB-A <35[36].9>, 36[37].2, 37[37].8, Pen.17 74[69].2, 74[69].5; *it*: RB-A 81[76].4

hidlir pres. ind. impers. < **hidlo*** *strain, filter, flow*: RB-A 38[38].4

hir adj. *long*: RB-A Add. 248v Bridging passage.5, Pen.17 109[103].3

hitheu conjunctive pron. 3 sg. f.: Pen.17 74[69].6, 74[69].8, 74[69].17, RB-A 74[69].15

holl[1] adj. *whole, all*: RB-A 5[5].1, 5[5].3, 6[6].1, RB-A 38[38].4 (twice)

holl[2] pron. *all, the whole*: RB-A 2[2].2, 2[2].7

holldedic verbal adj. *split/forked*: RB-A 12[13].3

honn dem. pron. f. *this*: RB-A 2[2].11, 27[29].2, 27[29].4, 30[32].2 <35[36].10>; **hon** RB-A 37[37].2

honno dem. pron. sg. f. *that*: RB-A 2[2].11, RB-B 2[2].2, RB-A 7[7].1, 8[8].2, <8[9].2>

hỽch n.m./f. *pig*: RB-A 12[13].18

hỽnn dem. pron. sg. m. *this*: RB-B Intro.1, RB-A 12[13].19, Pen.17 96[90].4, LH.21; **hunn** Pen.17 58[53].2; **hvnn** RB-A 81[76].6; see also **yr hỽnn**

hỽnnỽ dem. pron. sg. m. *that*: RB-A 2[2].2, 10[11].4, 10[11].6, 10[11].10 (twice); **hỽnnv** WB 32[33].8; **hvnnv** WB 32[33].10, 32[33].11, 32[33].12, 34[35].2, <35[36].2>; **hwnnỽ** RB-A 20[21].14, RB-B 34[35].2, 60[55].10; **hwnnu** Pen.17 71[66].2; **hvnnỽ** Pen.17 72[67].4

hỽy affixed pron. 3 pl. *they, them*: RB-A 37[37].5, RB-A Add. 248v Bridging Passage.6, RB-A LC.11

hynn pron. pl. *these, they*: RB-A 29[31].1, Pen.17 127[121].2; **hyn** Pen.17 73[68].6

hynny dem. pron. pl. *that, those*: RB-A 2[2].7, 2[2].13, 3[3].1, RB-B 3[3].5, RB-A 6[6].8; **henne** Pen.17 57[52].2, 60[55].14, 85[80].1, 86[81].3; **rei hynny** pron. *those ones*: RB-A 3[3].2, 10[11].12, 10[11].17, 12[13].23, RB-B 32[33].2; **rei henne** Pen.17 60[55].3, <61[56].1>, 73[68].2, 94[88].3

hynny dem. pron. sg. neut. *that*: RB-B Intro.1, RB-A 20[21].6, WB 23[24].5, RB-A 29[31].7, WB 29[31].5; **hyny** Ph. 23[24].5; **henne** Pen.17 71[66].3, 71[66].4, 74[69].6, 86[81].3, 87[82].1

hynt n.m./f. *path, way*: RB-A 9[10].12, Pen.17 96[90].1

hyt[1] conj. *until/as long as*: RB-A 2[2].11

hyt[2] n.m./f. *height, length*: RB-A 10[11].15, 10[11].19, 12[13].19, 12[13].22, 12[13].23

hyt[3] prep. *until, as far as, extending to*: RB-A 7[7].3, 7[7].4, 8[8].6, 10[11].3, 10[11].14, 12[13].4

hyt ar prep. *until, as far as*: RB-A 7[7].3, 32[33].9, Pen.17 88[83].4, 88[83].5, 88[83].6; **hit ar** Pen.17 88[83].3

hyt att prep. *as far as/to*: RB-A Add. 248v 88[83].3, 88[83].4, 88[83].7, 88[83].8, 88[83].9

hyt parth a'r prep. phrase + def. art. *towards the, as far as the*: RB-A 30[32].1–2

hyt yn prep. phrase *as far as, up to*: RB-A 9[10].8, WB 21[22].3, Ph. 21[22].3, RB-A 23[24].1; **hyt ym** (< yn before m-) RB-B 23[24].1, RB-A 24[25].3; **hyt y(m-)** WB 23[24].1, Phil. 23[24].1, RB-B 27[29].2, WB 27[29].2, RB-A 28[30].1; **hyt yg** (before k-) RB-B 21[22].3, Pen.17 86[81].5

i

i affixed pers. pron. 1 sg. *I*: 2[2].11, 2[2].13, Pen.17 71[66].3, RB-A LH.15, LH.16

iaỽn adj. *well*: RB-A 2[2].8; see also **yaỽn**

Idra *Hydra* (constellation): Pen.17 128[122].1

iechyt n.m./f. *health*: RB-A LH.2

ile (= Greek ὕλη *raw material*) *element*: RB-B 3[3].1

issel adj. *low*: Pen.17 60[55].10, RB-A 60[55].9

issaf adj. spv. *lowest* < is adj. *low*: RB-A 3[3].10, RB-B 3[3].13

Iupiter *Jupiter* (planet): Pen.17 86[81].2; **Jupiter** Pen.17 88[83].6; **Jouen** Add. RBA 248v 88[83].6 (twice)

k see c

ll

llad vn *kill*: RB-A 12[13].21; pres. ind. 3 sg. **llad** RB-A 29[31].3, RB-B 29[31].2, WB 29[31].2, RB-A <35[36].2>, <35[36].3>; pres. ind. 3 pl. **lladant** RB-A 10[11].22; pret. 3 sg. **lladaud** Pen.17 128[122].1, 140[134].2

llall pron. *the other/corresponding one*: RB-B Intro.2, RB-A 12[13].7, 12[13].8, 94[88].2; pl. **lleill** Pen.17 59[54].3, 74[69].2, WB 74[69].2

llauur n.m. *labour, work*: RB-A LH.8, LH.14, LH.15, LH.19

llafurio* vn *work*: **llauuryaf** pres. ind. 1 sg. RB-A 2[2].12; **llauurya** pres. ind. 3 sg. RB-A 2[2].11; **llaỽurya** Pen.17 74[69].16; **llafurya** RB-A 74[69].14; impf. ind. 3 pl. **lauuryynt** RB-A Add. 248v Bridging passage.5; pres. subj. 3 sg. **llauuryo** RB-A 42[42].1; impf. subj. 3 sg. **llaỽuryei** Pen.17 73[68].5

llauuryus adj. *assiduous*: RB-A LH.9

llanỽ n.m. *tide, flow*: RB-A 40[40].1, 40[40].3 (twice), 40[40].4, 40[40].5

llassowot pl. < **llysywen*** n.f. *eel*: RB-A 12[13].21

llau hep lau *mixed*: Pen.17 56[51].4

llaỽenhau vn *to rejoice*: RB-A LC.4

llawer pron. *many, large number, abundance*: RB-A 9[10].14, 10[11].13, 20[21].13, RB-B 32[33].6, WB 32[33].7; **llaỽer** RB-A 60[55].6, WB 60[55].3, RB-A 74[69].6; **llaver** Rawl. 60[55].6

llaỽn adj. *full*: RB-A 37[37].9

llaỽnlloer n.f. *full moon*: RB A-516 74[69].5

llaỽnlloneit n.m. *full* (of the moon), *full moon*: Pen.17 74[69].20, RB-A 74[69].19

lle n.m. *place*: RB-A 3[3].11, RB-B 3[3].14, RB-A 5[5].6, 6[6].5, 7[7].4; pl. **lleoed** RB-A 32[33].4, RB-B 32[33].6, WB 32[33].6, RB-A 37[37].10, 37[37].14, 37[37].16

lledu* vn *widen*: pres. ind. 3 sg. **lletta** RB-B <35[36].3>; **lleta** WB <35[36].3>

llef n.m/f. *shout* (here *musical note*): Pen.17 87[82].2

llehau* vn *place, set*: pret. impers. **llehawyt** RB-B 3[3].16; **llehaut** Pen.17 128[122].2

lles n.m. *benefit/advantage*: RB-A <35[36].15>

llestreit n.m. *vessel-ful/cask-ful*: Pen.17 63[58].3

llestri n. pl. < llestr* n.m. *vessel*; llestri dineu *pouring-vessels*: RB-A 33[34].12

llet adj. *broader*: WB 74[69].6

lleỽ n.m. *lion*: RB-A 12[13].3, 12[13].13; llew Pen.17 140[134].1; pl. llewot RB-A 10[11].20

lleuuer n.m. *light/resplendence*: Pen.17 67[62].2, 74[69].4, 143[136].2; lleuỽer Pen.17 74[69].5; lleuver RB-B 74[69].3; lleueuer WB 74[69].4

lleuuerhaa pres. ind. 3 sg. < lleuferu* *illuminate/shine*: RB-A 74[69].5

lleihau* *diminish*: pres. ind. impers. lleihaer RB-A LC.9; pres. ind. 3 sg. lleihaa Rawl. 60[55].7, Pen.17 74[69].12, RB-A 74[69].12

lleihaf adj. *least, smallest*: Pen.17 74[69].1, RB-B 74[69].1; lleiaf RB-B 60[55].8, WB 60[55].5, 74[69].1; byt lleihaf *microcosm*: 87[82].4; byt lleiaf 92[86].6

lleinỽ pres. ind. 3 sg. < llanw/llenwi* *flow, come in* (of the sea): RB-A 40[40].1, 40[40].2

lleuat n.f. *moon, Moon* (as one of the planets): RB-A 40[40].3 (thrice), 40[40].4, 40[40].6

lleygyon n. pl. < lleyg* n.m. *layman*: RB-A Add. 248v Bridging passage.6 (twice)

llin n.f. *line*: RB-A 6[6].7

llithraỽ vn *slide, flow, trickle*: RB-A 52[47].7, 52[47].8; pres. ind. 3 pl. llithrant RB-B 3[3].20, RB-A 9[10].12

lliw n.m. *colour*: RB-A 10[11].8, Pen.17 63[58].4 (twice), 63[58].5 (twice); lliỽ RB-A 12[13].14, 50[46].1, 74[69].17, RB A-516 74[69].3

lliwir pres. ind. impers. < lliwio/lliwo* *to colour, tint*: RB-A 10[11].8

lliwha pres. ind. 3 sg. < lliwhau* *to colour*: RB-A 50[46].2

lloer n.f. *Moon* (as one of the planets): RB-A Add. 248v 88[83].1, 88[83].2

llong* n.f. *ship*: n. pl. llongeu Pen.17 <61[56].3>; llogeu RB-A <61[56].3>

llosc n.m. *flaming heat*: RB-A 10[11].18

lloscỽrn n.m. *tail*: RB-A 12[13].6, 12[13].13, 12[13].18; lloscurn Pen.17 140[134].2

llosgi vn *burn, be on fire*: RB-A 10[11].22, 34[35].1, RB-B 34[35].2; llosci WB 34[35].2, RB-A 37[37].13; pres. subj. impers. lloscer RB-A 10[11].17

lluch n.m. *brightness/flame*: RB-A 60[55].6

lluesteu pl. < lluest* n.f. *dwelling (temporary)*: RB-A 12[13].25

llun see ar lun

llunyeith n.m./f. *arrangement/order*: RB-A LH.21; llunyaeth Pen.17 96[90].2

llunyethu vn *put in order/regulate*: Pen.17 97[91].1, Pen.17 146[139].2

llwydon pl. < llwyd* adj. *grey/pallid*: RB-A 11[12].5

llwytrew n.m. *hoarfrost*: Pen.17 67[62].3; llỽytreỽ RB-A 67[62].3, RB A-516 67[62].3

llydan adj. *broad/wide*: RB-A 12[13].4, 12[13].10

llyfyr n.m. *book*: RB-A LH.20; **llyuyr** RB-B Intro.1, RB-A 81[76].6, WB 81[76].3; pl. **llyfreu** RB-A Add. 248v Bridging passage.5, RB-A LH.20

llygeit n. pl. < **llygad*** n.m./f. *eye*: RB-A 11[12].10, 12[13].13

llyma adv. *here, behold, this*: RB-A 2[2].13

llyn n.m./f. *lake, pool*: RB-A 37[37].1; **llynn** RB-A 37[37].2

llyna adv. *there is, behold there*: RB-A 81[76].6, WB 81[76].2

llynedic adj. *defiled, polluted*: Pen.17 55[50].2

llyngu vn *swallow, devour*: Pen.17 109[103].3; pres. ind. 3 pl. **llyngkant** RB-A 12[13].1

llysseu collective n. *plants, vegetables, herbs*: RB-A <35[36].3>, RB-B <35[36].3>, WB <35[36].3>

llythyren n.f. *letter of the alphabet, character*: Pen.17 92[86].5

m

mab n.m. *son, youth*: RB-A 20[21].4; **map** Pen.17 109[103].2; **mab maeth** *foster son*: RB-A LH.1

maen n.m./f. *stone*: RB-A 33[34].10, 37[37].2; pl. **mein** Pen.17 65[60].2

magu* *grow, give birth*: pres. ind. 3 sg. **mac** RB-A 40[40].1, Pen.17 60[55].6, 60[55].7, 60[55].12, 60[55].14; pres. ind. 3 pl. **magant** RB-A 10[11].16, 11[12].5 (twice); pres. ind. impers. **megir** *are born*: RB-A 10[11].7, Pen.17 71[66].1, 72[67].1

Maharaen *Aries (sign of the Zodiac)*: Pen.17 98[92].1, 98[92].4; **maharaen** *ram*: Pen.17 98[92].2

mal prep. *like, as, similar*: RB-B <1[1].5>, <1[1].6>, RB-A 5[5].2, 5[5].5, 10[11].20; see also **ual**

mal yd conj. *according as*: RB-A 37[37].5, 52[47].2–3

malwot n. pl. *snails* (here *turtles*): RB-A 12[13].25

manac n.m./f. *information*: RB-A Add. 248v Bridging passage.7

manneu n. pl. < **man*** n.m./f. *blemish/stain*: Pen.17 74[69].19

manticora *manticore* (animal): RB-A 12[13].12

map see **mab**

march n.m. *horse*: RB-A 12[13].3, 12[13].5, 12[13].17; pl. **meirch** RB-A 12[13].16

marmor n.m. *marble*: RB-A 33[34].10

Mars *Mars*: Pen.17 86[81].2, 88[83].5; **Marten** RB-A Add. 247v 88[83].4, 88[83].5

marѷ adj. *dead*: RB-B <35[36].2>, RB-A 37[37].1; **marv** WB <35[36].2>; **meirѷ** pl. RB-A 11[12].15

maѷr adj. *great/big*: RB-A 12[13].3, 32[33].4, RB-B 32[33].6, RB-A <35[36].9>, LC.1; **mavr** WB 32[33].6

medeginyaeth n.m./f. *medicine, cure*: RB-A <35[36].4>, RB-B <35[36].4>, WB <35[36].4>

medỽl n.m. *mind, intent*: RB-A 2[2].2 (twice), 2[2].4, RB-B 2[2].1

megys prep. *like, as, as though, similar to*: RB-A 1[1].1, RB-B <1[1].2>, <1[1].3>, <1[1].4>, RB-A 3[3].12

megys y conj. *like, as*: RB-A 2[2].4, 36[37].1, Pen.17 98[92].3, RB-A LC.10; megys na Pen.17 95[89]2

meheuin n.m. *June*: RB-A <35[36].14>

mein pres. ind. 3 sg. < manu* *prevail, avail*: RB-A 12[13].11

meint n.m./f. *extent, dimension*: RB-A 12[13].9, 37[37].13

mellt pl. n. *flashes of lightning*: Pen.17 60[55].8, 60[55].13, RB-B 60[55].6, 60[55].10, WB 60[55].4

melyn n.m. *yolk of egg* (lit. *yellow*): RB-A 1[1].4 (twice), 1[1].7, RB-B <1[1].3>, <1[1].4>

melys adj. *fresh* (of water): RB-A 52[47].1, 52[47].2; cpv. melyssach RB-A 52[47].9

melynyon pl. adj. < melyn* *yellow*: RB-A 12[13].9

menegi vn *narrate, explain*: RB-A Add. 248v Bridging passage.6; pres. ipv. 1 pl. mynagỽn RB-A 32[33].10

mer n.m. *essence*: RB-A LH.4

merch n.m. *daughter*: RB-A 21[22].1

Mercurius *Mercury*: Pen.17 86[81].1, 86[81].4, 88[83].2; Mercuriỽm RB-A Add. 248v 88[83].2

mererit n.m./n. pl. *pearls*: RB-A LH.15

meridies Lat. *south*: RB-A 7[7].4; meridiem RB-A 7[7].3; see also hanner dyd

messur n.m. *measure*: Pen.17 86[81].3

mesur* vn *measure*: pres. ind. impers. messurir Pen.17 85[80].4; pret. impers. messurỽyt RB-A 5[5].1

meuyryant pres. ind. 3 pl. < myfyrio* *meditate (upon)*: RB-A LH.13

mi pers. pron. 1 sg. *me, I, myself*: RB-A LC.6, LH.6

mil num. *thousand*: RB-A 5[5].1, Pen.17 86[81].6, 88[83].2, 88[83].3, 88[83].5; pl. milioed RB-A 5[5].2, RB-A Add. 248v, 88[83].9, 88[83].10; milyoed Pen.17 86[81].6, 88[83].2, 88[83].3, 88[83].5, 88[83].6

milltir n.m./f. *mile*: RB-A 5[5].1, Pen.17 86[81].6, 88[83].1, 88[83].2, 88[83].4; pl. milltiroed RB-A 5[5].2, Pen.17 86[81].6, 88[83].2, 88[83].3, 88[83].7

minheu conjunctive pron. 1 sg.: RB-A LH.18; minnheu RB-A 2[2].11

mis n.m. *month*: RB-A 29[31].5 (twice), RB-B 29[31].4, WB 29[31].4 (twice)

miui reduplicative pers. pron. 1 sg. *me*: RB-A LH.9, LH.18

moch pl. n. *swine*: RB-A LH.15

mod n.m. *way, manner*: RB-A 2[2].1

moelronyeit n. pl. < moelrhon* n.m./f. *hippopotamus*: Pen.17 56[51].6

molesteu n. pl. < molest* n.m./f. *worry*: RB-A LH.10

monochero *monoceros* (*unicorn*): RB-A 12[13].17

mor n.m. *sea*: RB-A 5[5].5, 9[10].5, 10[11].4, 10[11].6, 10[11].10; pl. **moroed** RB-A 34[35].2, 38[38].1, 52[47].6

moraỽl adj. *marine*: RB-A 34[35].3

morduywyr n. pl. < **mordwywr*** n.m. *seaman/sailor*: Pen.17 57[52].1, 96[90].1

morgerỽyn n.f. *whirlpool, vortex, abyss*: RB-A 37[37].10

morỽoch pl. n. *dolphins*: Pen.17 57[52].2

mỽc n.m. *smoke*: RB-A 37[37].4, 37[37].13

mur n.m./f. *wall*: RB-A 8[8].6, <31[32].2>

mussic n.m./f. *music*: Pen.17 86[81].3, 87[82].4

mỽy cpv. adj. *more, bigger*: RB-A 5[5].4, <35[36].10>, 74[69].1, 74[69].8 (twice); **mwy** Pen.17 74[69].2, 74[69].8, 74[69].9; **muy** Pen.17 74[69].9; **mvy** WB 74[69].2; spv. **mỽyhaf** *greatest*: RB-A 32[33].4, 60[55].8, LH.23; **muyhaf** Pen.17 60[55].10

muyhaa pres. ind. 3 sg. < **mwyhau*** *grow*: Pen.17 74[69].12

myhun pers. pron. 1 sg. + **hun** *myself*: RB-A LH.15, LH.16

mynet *go*: RB-B 32[33].3, WB 32[33].4; pres. ind. 3 sg. a RB-A 5[5].4, 9[10].8, 10[11].2, RB-B 23[24].3, WB 23[24].3, 23[24].4; **aa** RB-B 23[24].4; pres. ind. 3 pl. **ant** RB-A <8[9].6>, <8[9].7>, 10[11].22, 52[47].5; ipv. 1 pl. **aỽn** RB-A 29[31].8; pres. subj. 3 sg. **el** RB-A 40[40].3, 40[40].5, 48[45].1, 49[45].1

mynnỽyt vb pret. impers. < **mynnu*** *want, obtain*: RB-B 2[2].5

mynyd n.m. *mountain*: RB-A 10[11].10, RB-B 32[33].11, 32[33].12, WB 32[33].11, 32[33].12; pl. **mynyded** RB-A 10[11].8, 10[11].14, 10[11].17, RB-B 21[22].2, WB 21[22].2

myỽn prep. *within*: RB-A <8[9].6>, 37[37].9; **y myỽn** *in, inside*: RB-A LC.10, LH.22

n

na(t) see **ny**

na... na *neither... nor*: RB-A 12[13].11, <35[36].1>, RB-B <35[36].1>, WB<35[36].1>, RB-A Add. 248v Bridging passage.4–5

namyn prep. *except, but, but rather*: RB-A 5[5].4, 6[6].10, 11[12].13, RB-B 29[31].5, WB 29[31].5

namyn conj. *except, but*: RB-A 52[47].6, 74[69].4, RB-B 74[69].3, WB 74[69].4; **namen** Pen.17 55[50].4, 74[69].4, 74[69].13, 95[89].4

naỽ num. *nine*: RB-A 34[35].4, RB-B 74[69].5, WB 74[69].6, Pen.17 86[81].8, RB-A Add. 248v 88[83].10; **nau** Pen.17 86[81].8, 88[83].10, 146[139].2

neb pron. *someone/no one*: RB-B 32[33].3, 32[33].7, WB 32[33].3, 32[33].4, 32[33].7; **y neb** rel. antecedent *the one*: RB-A 2[2].4

neb ryỽ pronominal *any, any kind of*: RB-A <35[36].15> (twice), <35[36].16>

nef n.m./f. *sky, heaven*: RB-B <1[1].4>, <1[1].5>, RB-A 8[8].7, 32[33].9, Pen.17 86[81].8

neidyr n.f. *snake, serpent*: RB-A 12[13].14, <35[36].1>; pl. **nadred** WB 29[31].3; **anatred** RB-B 29[31].3

neges n.m./f. *matter, business*: RB-A LH.8

(y) neill pron. *one, this*: RB-A 12[13].7; **neilltu** n.m. *one side*: Pen.17 94[88].1; **o'r neilltu** adverbial phrase *on the one side*: RB-A 6[6].7; **o'r neill parth** adverbial phrase *on the one side*: RB-A 6[6].3

nerth n.m./f. *strength, virtue*: RB-A 10[11].16, Pen.17 87[82].3

nessaf adj. *nearest, next (to)* < **agos*** adj. *near*: RB-A 3[3].13 (twice) RB-B 3[3].15, 20[21].4, 20[21].5

nesset abstract noun *nearness*: RB-A 40[40].4, 74[69].8, WB 74[69].2

neu conj. *or, or rather*: RB-A 21[22].1, 37[37].1, 37[37].5, 38[38].3, 39[39].1

neuaul adj. *heavenly*: Pen.17 85[80].4, 87[82].5, 88[83].12

newyd see **o newyd**

ni¹ affixed pers. pron. 1 pl. *we, us*: Pen.17 85[80].2, 85[80].3, 94[88].2, RB-A LC.11

ni² affixed poss. pron.1 pl. *our*: Pen.17 85[80].3, RB-A LH.14, LH.22

ninheu pers. pron. conjunctive 1 pl. *we*: Pen.17 85[80].5, 87[82].2, 87[82].3, 94[88].2, 95[89].3

no conj. *than*: RB-A 29[31].4, 49[45].2, Pen.17 74[69].9, RB-A 74[69].9, RB-A LH.9; **noc** RB-A 5[5].4, 11[12].14, 12[13].15, Pen.17 85[80].2, RB-A LH.16; with def. art. **no'r** RB-A 11[12].8, 32[33].3, <35[36].10>, 52[47].9, Pen.17 72[67].3

noeth adj. *pure, bare*: RB-A 6[6].4

nofio* *swim*: pres. ind. 3 sg. **nouya** Pen.17 58[53].3; pres. ind. 3 pl. **nofyant** RB-B 3[3].18, RB-A 12[13].1

nos n.f. *night*: RB-A 29[31].6 (twice), RB-B 29[31].4, WB 29[31].4, RB-B 32[33].3

Nothus (wind): RB-A 60[55].6, RB-B 60[55].5, WB 60[55].3, Rawl. 60[55].5

ny neg. part. *not*: RB-A 6[6].1, 6[6].4, 10[11].9, 10[11].24, 11[12].6; **nyt** RB-A 5[5].3, 5[5].4, 6[6].10, 11[12].13, 29[31].4; **na** RB-B 32[33].3 (twice), WB 32[33].3, 32[33].4, RB-A 52[47].6; **nat** 6[6].5; with infixed obj. pron. sg. **nys** RB-B, 32[33].6, WB 32[33].7; **nyt. amgen** grammaticalized phrase *not otherwise, namely*: RB-A 3[3].2, 3[3].11, RB-B 3[3].5, RB-A 29[31].4, 30[32].7; **nyt dim amgen** *not anything other/else*: Pen.17 59[54].2; **ny... nac ny...** *neither (vb)... nor (+ vb)...* RB-A 74[69].11–12

ny neg. rel. part.: Pen.17 94[88].2

nyt 3 sg. pres. neg. of copula: RB-A 38[38].5, RB-A Add. 248v Bridging passage.4, RB-A LH.18, LH.20

nyⱱl n.m. *fog, mist, clouds*: RB-B 60[55].5, WB 60[55].2, Rawl. 60[55].5

o

o¹ prep. *from, of, on account of*: RB-A 1[1].1, 1[1].3, RB-B intro.2, <1[1].1>, RB-A 2[2].1; *or* RB-A 12[13].21; with suffixed personal pron. 2 sg. **ohonot** RB-A

LC.9; 3 sg. m. **ohonaỽ** RB-B 3[3].2, RB-A <8[9].2>, Pen.17 92[86].1; **ohonav** Pen.17 55[50].3, 59[54].1; 3 sg. f. **ohonei** RB-A 21[22].2, 30[32].2, 50[46].1, Pen.17 74[69].4, 74[69].16; 3 pl. **onadunt** RB-A 2[2].1, 3[3].5, 3[3].10, 6[6].1, 9[10].1; **ohonunt** RB-A 52[47].4; with def. art **o'r** *of the, from the*: RB-B <1[1].2>, RB-A 2[2].10, 3[3].12, RB-B 3[3].11, RB-A 5[5].4; with dem. art. **ar** as **o'r** Pen.17 58[53].1, RB-A 74[69].13, RB A-516 74[69].3, Pen.17 85[80].3, RB-A LC.10; with poss. pron. 2 sg. **o'th** *of your*: RB-A LC.9; with poss. pron. 3 sg. m. **o'e** RB-B 2[2].4, RB-A 12[13].18, 37[37].7, 37[37].9, Pen.17 55[50].4; with pron. 3 sg. f. **o'e** Pen.17 73[68].5, 74[69].14, 87[82].5, 93[87].1, 93[87].2

o² marks agent of vn: RB-A 10[11].11, RB-A Add. 248v Bridging passage.6

o³ conj. *if*: Pen.17 74[69].18; with affirmative particle **ry**: **o'r** RB-A 12[13].11

o achaỽs see **achaus**

o hynny *from that, on that account*: RB-A 36[37].2; **o henne** Pen.17 56[51].1, 86[81].3, 86[81].7, 87[82].4

o'e myỽn prep. with obj. pron. 3 sg. f. *from inside it* (< **o fewn*** *within, from inside*): RB-A 37[37].9

o newyd prep. *anew*: RB-A 2[2].13, 9[10].9, 40[40].6–7

oc prep. *of, from*: RB-A 3[3].5, 10[11].23, 11[12].9, 12[13].25, 33[34].1

oduch prep. *above*: RB-A Add. 248v Bridging passage.1

odyma adv. *from here*: RB-A 74[69].9

odyna adv. *thence, afterwards*: RB-A 1[1].5, 1[1].6, 5[5].5, 10[11].1, 20[21].2; **odena** Pen.17 86[81].1, 86[81].4, 86[81].5, 88[83].3, 88[83].4

odyno adv. *thence*: RB-A 33[34].11, 42[42].2

oed n.m./f. *age*: RB-B 2[2].3

oer¹ adj. *cold*: RB-A 3[3].7, 3[3].8 (twice), RB-B 3[3].8, 3[3].9; eq. **oeret** RB-B 32[33].3, WB 32[33].3

oer² n.m. *cold*: RB-B 3[3].8, RB-A 6[6].6

oeruel n.m./f. *cold*: RB-A 6[6].2, 6[6].3, 6[6].4, 6[6].7, 29[31].7, 37[37].15; **oerỽel** Pen.17 65[60].1

oeruelauc adj. *cold*: Pen.17 81[76].1

oessoed n. pl. < **oes*** n.f. *period of time, age*: RB-A 2[2].2, LC.3

ogylchynu vn *perambulate/surround*: RB-A 9[10].9

oll pron. *all*: Pen.17 71[66].3, 81[76].4, RB-A Add. 248v Bridging passage.4, Pen.17 93[87].3, RB-A LH.15, LH.16

olwyn n.m./f. *wheel*: Pen.17 73[68].6

onyt conj. *if… not*: RB-A LH.23

o'r see **o¹** and **o³**

or conj. *if*: RB-A 12[13].21, <35[36].16>

o'r tu arall see **tu**

o'r tu hỽnt (y) see **tu**

Orcius (wind) WB 60[55].2; see also **Circius**

organ n.m./f. *musical* (esp. wind) *instrument*: Pen.17 87[82].5

os *if it be* (< **o** conj. *if* + **ys** pres. ind. 3 sg. of **bot**): Pen.17 67[62].2, RB-A 67[62].2, Pen.17 74[69].19, 74[69].20, RB-A 74[69].16

P

pa... bynnac *whatsoever*: RB-A 29[31].2

pa du bynnac pron. *wherever*: RB-A 5[5].7

paham pronominal *why, wherefore, how*: Pen.17 55[50].1, 56[51].4

palader n.m./f. *spear, shaft*: Pen.17 63[58].1, 63[58].2

palestyr (herb) *parsley?*: RB-B <35[36].3>, WB <35[36].3>

pan conj. *when*: RB-A 2[2].4, 2[2].5, 2[2].8, 2[2].12, RB-B 2[2].3; **pann** RB-B 2[2].2, RB-A 10[11].17, 12[13].6, 27[29].3, RB A-516 67[62].1; **pan yỽ** *that it is*: RB-A 52[47].4, <61[56].3>

Paraduys sg. *Paradise* Pen.17 146[139].2; pl. **Paraduysseu** Pen.17 146[139].3

parant pres. ind. 3 sg. < **peri*** *cause*: RB-A 42[42].3

parhau* *remain, continue, last, survive*: pres. ind. 3 sg. **para** RB-A <8[9].3>, 49[45].1; pres. ind. 3 pl. **parhaant** RB-A 11[12].6

parth a('r) prepositional phrase *towards (the)*: RB-B 27[29].1, WB 27[29].1, RB-B 28[30].1, WB 28[30].1, RB-A 29[31].1

paỽp pron. *each, every* (here *everyone, all*): RB-A 8[8].5, <35[36].16>; **paỽb** RB-A 81[76].4; **paub** Pen.17 58[53].2

pedeir num. f. *four*: RB-A <8[9].5>, 10[11].13, 12[13].18, 33[34].4, 52[47].2, 52[47].3

pedwar num. m. *four*: RB-A 3[3].1, RB-B 3[3].3, RB-A Add. 248v 88[83].8; **petwar** Pen.17 59[54].3, 60[55].1, 87[82].3, 88[83].5, RB-A Add. 248v 88[83].4

pedweryd num. m. *fourth*: RB-B 2[2].3, RB-A 60[55].9, RB-B 60[55].8; **pedwyred** RB-A 6[6].9; **petweryd** Pen.17 60[55].10, WB 60[55].6, Pen.17 72[67].1, 86[81].2; **pedwyryd** RB-A 74[69].16, RB A-516 74[69].3; **pydỽryd** Rawl. 60[55].8; f. **pedwared** RB-A 2[2].8

pei conj. *if*: RB-A 6[6].4, Pen.17 63[58].3, 73[68].4

pei na bei *except*: Pen.17 74[69].7, RB-A 74[69].7, RB-B 74[69].4, WB 74[69].5

peidyei impf. ind. 3 sg. < **peidio*** *stop/rest*: Pen.17 109[103].3

pel n.f. *ball*: RB-A 1[1].2, RB-B <1[1].2>

pell adj. *far, distant, far-off*: RB-A <35[36].15>; eq. **pellet** (used as n.) *far, remote*: RB-A 11[12].13, 40[40].5; spv. **pellaf** Pen.17 74[69].11, RB-A 74[69].11, RB A-516 74[69].2; **o bell** adv. *from afar* RB-A 6[6].5; with nominal meaning *in the distance* **ym pell**: RB-A <8[9].7>; **em pell** Pen.17 147[140].1

pellen n.f. *globe, round object*: Pen.17 74[69].3; **pellenn** RB-A 74[69].3

penn n.m. *head*: RB-A 11[12].2, 11[12].10, 12[13].9, 12[13].17, Pen.17 128[122].1

pennaf adj. *chief, principle*: RB-A LH.16

perffeith adj. *perfect*: RB-A Add. 248v Bridging passage.6

periglus adj. *dangerous*: RB-A LH.10

perigyl n.m. *danger, risk*: RB-A LH.9; pl. **perigleu** RB-A 34[35].2, RB-B 34[35].3; **periglev** WB 34[35].2

perued n.m. *middle, centre*: RB-A 1[1].5, RB-B 3[3].15, RB-A 6[6].2, Pen.17 74[69].20, RB-A 74[69].18; **ym perued** *in the middle, centre (of)*: RB-A 12[13].18, 33[34].7–8, 36[37].1, 36[37].1–2, Pen.17 93[87].1

peth n.m./f. *thing, object*: RB-A 2[2].3, RB-A Add. 248v Bridging passage.4; pl. **petheu** RB-B 3[3].17, Pen.17 88[83].12, RB-A LH.11, LH.12, LH.13

petwarlliwyauc adj. *four-coloured*: Pen.17 63[58].1

petweryd see **pedwared**

peunyd adv. *daily, always, continually*: RB-A 40[40].2, Pen.17 74[69].14, RB-A 74[69].13

Pistrix *Pistrix* (constellation): Pen.17 140[134].1

planet n.m./f. *planet*: RB-A 74[69].1, RB-B 74[69].1, WB 74[69].1, Pen.17 81[76].1, RB-A 81[76].1; pl. **planedeu** Pen.17 73[68].2, 73[68].6, 74[69].1, 88[83].11; **plannedeu** RB-A 81[76].2, RB-A Add. 248v Bridging passage.2

plisgyn n.m. *shell* < pl. n. **plisg***: RB-A 1[1].3

plith dra chemysc *in confusion, mixed (up)*: Pen.17 56[51].3–4

pob pron. *each, every*: RB-A 5[5].6; **pop** RB-A 11[12].5, <35[36].15>; **pob amser** *always*: RB-A 10[11].6; **pob kyfryѵ** *all kinds of, every kind*: RB-A 8[8].4; **pob eilwers** *one by one, alternately, in turns*: RB-A 3[3].7; **pop eilwers** RB-A 12[13].10–11; **pop parth** *every side*: RB-A 1[1].1, 6[6].5, 6[6].7, 20[21].8; **pob rei** *each one*: RB-A 3[3].5, 38[38].4, RB-A Add. 248v Bridging passage.3; **pob ryѵ** *every kind of*: RB-A LH.2; see also **pob peth**

pob un pron. *everyone, all*: RB-A 2[2].9 (twice), 3[3].2, RB-B 3[3].7, RB-A 10[11].22; **pob vn** RB-B 2[2].4, RB-A 3[3].6; **pop vn** Pen.17 59[54].3

pob peth pron. *everything*: RB-A 2[2].5, RB-B 3[3].4, RB-A 10[11].5–6, Pen.17 58[53].1, 60[55].6; **pob ppeth** RB-A 2[2].6; **pop peth** RB-A 2[2].13, 3[3].1–2

poen n.m./f. *pain, torment*: Pen.17 58[53].4; pl. **poeneu** RB-A 37[37].14

poenaѵl adj. *painful, penal, tormenting*: RB-A 37[37].14

popyl n.m./f. *people, nation, tribe*: RB-A 11[12].1; **pobyl** RB-A 11[12].4, 11[12].7, 23[24].2, WB 23[24].2, Ph. 23[24].2; pl. **poploed** RB-A 10[11].14; **pobloed** RB-A 10[11].18, 10[11].21, RB-B 23[24].1, 32[33].4, WB 32[33].5

porffor n.m. *purple colour*: RB-A <35[36].7>

prenn n.m. *tree*: RB-B 2[2].4 (twice), RB-A <8[9].1>, <35[36].11>

pressѵylaѵ *abide, live (in), inhabit*: RB-A 6[6].2; **pressѵylaѵ** RB-A 6[6].5; **pressuyllyau** Pen.17 56[51].5; pres. ind. 3 sg. **pressѵyla** RB-A 2[2].5; **pressuylua** Pen.17 146[139].2; **presuyllya** 147[140].1; pres. ind. 3 pl. **pressuyllyant** Pen.17 56[51].1, 56[51].2

pressỽyledic adj. *inhabitable*: RB-A 6[6].10

pressỽyluodaf n. with spv. ending < **preswylfod*** *dwelling place* (rendering Lat. *inhabitatione gloriosa*): RB-A 20[21].12

priaut adj. *proper, specific, individual*: Pen.17 59[54].3, 73[68].5, 74[69].4, WB 74[69].3; **priaỽt** RB-A 74[69].4, RB-B 74[69].3

prid n.m. *soil*: RB-B 29[31].2, WB 29[31].2

prif n. *primary*: RB-B 23[24].3, WB 23[24].3, Ph. 23[24].3; *prime (of the moon)*: RB-A 74[69].17, RB A-516 74[69].3

prifwynt *cardinal/principal wind*: Pen.17 60[55].1, 60[55].4, 60[55].7, 60[55].10, RB-A 60[55].3; pl. **prifwynhyeu** Pen.17 59[54].3

priodolder n.m./f. *property, attribute*: RB-A 3[3].6

provi vn *prove*: RB-B 81[76].3; **proui** WB 81[76].2; pres. ind. impers. **prouir** Pen.17 73[68].4; impf. impers. ind. **prouit** Pen.17 81[76].3

pryt na conj. + neg. *for... not...*: RB-A LC.9, RB-A LH.16

pryuet n. pl. < **pryf*** n.m. *reptile, maggot, wild animal*: RB-A 12[13].22, 29[31].3, 37[37].9

pur awyr *ether*: Pen.17 72[67].4; **aỽyr pur** RB-B <1[1].5>

p6nc n.m. *point (in space), pole*: RB-A 40[40].5

puy bennac pron. wiwhowhoever*whosoever/whatsoever*: Pen.17 81[76].2; **pỽy bynnac** RB-A 81[76].4, RB-B 81[76].1; **pwy bynnac** WB 81[76].1

pỽynt n.m./f. *point*: RB-A 5[5].2, 5[5].3

pybyr n.m. *pepper*: RB-A 10[11].17

pyc n.m. *bitumen, pitch*: Pen.17 55[50].2 (twice), 55[50].3, 55[50].4

pydeỽ n.m. *well, pit, mire*: RB-A <35[36].12>

pylo pres. subj. 3 sg. < **pylu*** *become blunt*: RB-A 12[13].7

pym num. *five*: RB-A 2[2].1, 30[32].5, 39[39].1; **pum** RB-A 6[6].1; **pump** RB-B 2[2].1, RB-A 39[39].1, RB-A Add. 248v 88[83].5; **pump cant** *five hundred*: RB-A 81[76].5–6

pymhet num./adj. *fifth*: RB-A 2[2].12, RB-B 2[2].5, RB-A 6[6].10, 11[12].6; **pymet** Pen.17 86[81].2

pymthec num. *fifteen*: RB-A <31[32].3>, Pen.17 88[83].4, 88[83].8, RB-A Add. 428v 88[83].1, 88[83].4; **pymtheg** RB-A Add. 428v 88[83].1, 88[83].5; **pymtheng** Pen.17 86[81].6, 88[83].1, 88[83].10

pymthecant num. *fifteen hundred*: (**pymtheg** + **cant**) Pen.17 81[76].5

pyscaỽt n. pl. *fish*: RB-A 10[11].24; **pyscaut** Pen.17 56[51].1, 58[53].3, 109[103].2, 109[103].3; *Pisces (sign)* Pen.17 109[103].1

r

rac prep. *from, because of, on account of*: RB-A 6[6].2 (twice), 6[6].5, 9[10].14, 10[11].8

rac llaỽ adv. *henceforth*: RB-A LH.22–23

rann n.f. *part*: RB-A 6[6].1 (twice), 6[6].2, 7[7].1, 11[12].1 (twice); **ran** RB-A 7[7].1, Pen.17 56[51].6; pl. **ranneu** RB-A 6[6].8, 32[33].5, 32[33].9

redec vn *running, racing*: RB-A 12[13].15, RB-B 32[33].5, WB 32[33].6, Pen.17 73[68].2, 73[68].4

rei pron. subst. *some*: RB-A 74[69].2, LC.6; **pob rei** *each one*: RB-A 3[3].5, 38[38].4, RB-A Add. 248v Bridging passage.3; with def. art. as a rel. pron., antecedent to relative clause *the ones, those ones*: **y rei** RB-B 3[3].6, 3[3].8, 3[3].9 (twice), RB-A 20[21].2; **('r) rei hynn** *these (ones)* RB-B 3[3].12; **(y/yr/'r) rei hynny** pronominal use *those ones*: RB-A 3[3].2, 10[11].12, 10[11].16, 12[13].23, RB-B 32[33].2; **e/y/'r rei henne** Pen.17 60[55].2–3, <61[56].1>, 73[68].2, 94[88].3

rennir pres. ind. impers. < **rhannu*** *separate*: RB-A 6[6].1, 7[7].1, 38[38].2

rew n.m. *frost, ice*: Pen.17 60[55].4

rewedic verbal adj. *frozen*: RB-A 29[31].7

rewi vn *freeze*: Pen.17 65[60].1; pres. ind. 3 pl. **rewant** RB-A <61[56].2>; **rewhant** RB-B <61[56].2>; **revant** WB <61[56].2>

reỽlyt adj. *frozen*: RB-A 81[76].1

rieni pl. n. *parents*: RB-A 10[11].23

rif n.m. *sum, number*: Pen.17 96[90].3

rifir pres. ind. impers. < **rhifo*** *to count*: RB-A 3[3].14

rinwed n.m./f. *nature, quality*: Pen.17 96[90].2

rith n.m./f. *shape, form* (here *seed, embryo*): RB-B <1[1].4>, <1[1].6>

rithwt pret. impers. < **ritho*** *shape, transform*: Pen.17 109[103].3

riuedi n.m. *number(s), sum, total*: RB-A 5[5].1

rhoddi* *supply, give*: pres. ind. impers. **rodir** RB-A LH.21; pret. 3 sg. **rodes** Pen.17 127[121].1

rot n.f. *wheel*: Pen.17 94[88].3

rot melin *mill-wheel*: Pen.17 73[68].5

rⱱng see **y rⱱng**

rⱱym n.m. *bond*: RB-A 3[3].12

rⱱymedic verbal adj. *bound (up), fastened*: RB-B 60[55].7, Rawl. 60[55].7; **rwymedic** Pen.17 95[89].3

rwymo* *bind, restrain*: pres. ind. 3 sg. **rⱱym** RB-A 38[38].4, RB-B 38[38].3, WB 38[38].2, Rawl. 38[38].2; pres. ind. impers. **rⱱymir** RB-A 38[38].3

ry perfective particle used with vn: RB-A 10[11].11, Pen.17 57[52].3, 71[66].4, 88[83].11, 97[91].1

rydhaa pres. ind. 3 sg. < **rhyddhau*** *to free*: Pen.17 60[55].11

ryⱱ n.m./f. *kind, type, sort*: RB-A 2[2].6, RB-B 2[2].4 (m.), RB-A 11[12].1, 11[12].7, 52[47].1; **y ryⱱ** *that kind of*: RB-A 10[11].21, 11[12].4, 12[13].22, 12[13].24; **y ryu** Pen.17 56[51].4

S

Sadurn *Saturn* (planet): Pen.17 81[76].1, 81[76].3, 86[81].3, 88[83].8; **Saturnus** RB-A 81[76].1, 81[76].3, RB-B 81[76].1; **Saturniỽm** RB-A Add. 248v 88[83].7, 88[83].8

Sagitta Lat. *Sagitta* (constellation 'Arrow'): Pen.17 127[121].1

sardinus *sardius* (precious stone): RB-A 33[34].10

sarff n.m./f. *snake*: RB-A 12[13].13, RB-B <35[36].3>, Pen.17 140[134].2; **sarph** WB <35[36].1>; pl. **seirff** RB-A 9[10].14, 10[11].9, 10[11].18, 12[13].1, RB-B 32[33].7; **seirf** RB-A <35[36].6>; **sarphot** WB 32[33].7

sathru vn *trample, walk on*: RB-A LH.12; pres. ind. 3 pl. **sathrant** RB-A LH.14

sech adj.f. *dry*: RB-A 3[3].7, 3[3].9; see also **sych**

sef[1] pron. *that is*: RB-B intro.1, RB-A 10[11].12, Pen.17 60[55].7, 60[55].11, <61[56].3>

sef[2] pron., predicative use: RB-A <35[36].14>

sef[3] particle before rel. clause: Pen.17 74[69].1, RB-A 74[69].1

seith num. *seven*: RB-A 9[10].10, 11[12].8, 23[24].5, RB-B 23[24].3, WB 23[24].4; **seith naỽn yr Yspryt Glan** *seven gifts of the Holy Spirit*: RB-A LC.3

seithuet num. *seventh*: Pen.17 81[76].1, RB-A 81[76].1, RB-B 81[76].1, Pen.17 86[81].3

semis Lat. *semis* n. sg.m. *half*: Pen.17 88[83].3

Septemtrio Lat. *north* (one of the cardinal directions), *north wind*: RB-A 7[7].3, 10[11].2, 60[55].1, WB 60[55].1, Pen.17 92[86].3; **Septentrio** Pen.17 60[55].1, RB-B 60[55].1, Rawl. 60[55].1

septemtrionalis Lat. adj. nominative sg. *North Frigid Zone*: RB-A 6[6].9

seren see **syr**

seucocreta *Leucocrota* (animal species): RB-A 12[13].2

soddi* *sink, submerge*: pres. ind. 3 sg. **saỽd** RB-A 12[13].24, 37[37].2, 37[37].3; pret. 3 sg. **sodes** RB-A <35[36].9>

solifuga (type of animal): RB-A <35[36].2>

solsticialis Latin adj. nominative sg. *North Temperate Zone*: RB-A 6[6].9

solsticỽm *solstice*: RB-A 29[31].3

Subsolanus *Subsolanus* (wind): Pen.17 60[55].4, RB-A 60[55].3, RB-B 60[55].3, Rawl. 60[55].3

sugnaỽ vn *to drink, absorb, attract*: RB-A LH.4

sych adj. m. *dry*: RB-A 3[3].9, RB-B 3[3].7, 3[3].11 (twice); see also **sech**

sycha pres. ind. 3 sg. < **sychu*** *to (make) dry*: Pen.17 60[55].6, RB-A 60[55].5, RB-B 60[55].4, Rawl. 60[55].4

sychdỽr n.m. *dryness*: RB-A 5[5].6

sychedic verbal adj. *parched*: RB-A LH.4

sychin n.f. *dry weather, draught*: RB-A 74[69].18; **sychhin** RB A-516 74[69].5

sygyn n.f. *sign* (of the Zodiac): Pen.17 81[76].2, 97[91].1, 98[92].1, 98[92].3, 98[92].5;

pl. **sygneu** Pen.17 73[68].6, 74[69].16, 86[81].5, RB-A Add.248v Bridging passage.1, Pen.17 96[90].3

symudaỽ vn *transform*: RB-A 12[13].14, Pen.17 65[60].2

syr pl. n. *stars*: Pen.17 74[69].1, WB 74[69].1, RB-A Add.248v Bridging passage.2, Pen.17 93[87].2, 95[89].1; **ser** RB-B 32[33].12, WB 32[33].12, RB-B 74[69].1; sg. **seren** n.f. Pen.17 73[68].1, 92[86].4

t

tadogaeth n.f. *paternity, origin, pattern*: Pen.17 85[80].4

taflu vn *throw, surge* (of flames): RB-A 37[37].11; pres. ind. 3 sg. **taflo** RB-A 40[40].1

tal n.m. *forehead*: RB-A 12[13].18

tan n.m. *fire*: RB-A 3[3].2, 3[3].3, 3[3].5, 3[3].9, 3[3].10

tanaỽl adj. *fiery*: RB-A 37[37].9, 74[69].3; **tanaul** Pen.17 63[58].4, 74[69].3, 96[90].2

taraneu n. pl. < **taran*** n.f. *thunder*: Pen.17 60[55].13, RB-A 60[55].10, Rawl. 60[55].10; **taranev** WB 60[55].7

Tat n.m. *Father*: 2[2].11; **Tat Trindaỽt** n.m. *Father of the Trinity*: RB-A LC.5

tebic adj. *resembling*: RB-B <35[36].3>, WB <35[36].3>

tegỽch n.f. *beauty, fairness*: RB-A 8[8].5, <35[36].15>

teir num. f. *three*: RB-A 7[7].1, 12[13].13, RB-A Add. 248v 88[83].3, 88[83].8

teirỽ n. pl. < **tarw*** n.m. *bull*: RB-A 12[13].9

temystyl see **tymestyl**

teneuach adj. cpv. < **tenau** adj. *thin, flat*: Pen.17 72[67].2, 72[67].3 (twice)

teneuheir pres. ind. impers. < **teneuhau*** *become rare, thin out*: RB-A 38[38].3

teruyn n.m. *boundary, limit*: RB-A 5[5].4, 7[7].2; **terỽyn** Pen.17 72[67].1; pl. **teruyneu** RB-A 10[11].3

terfynu* *end, conclude, terminate*: pres. ind. 3 sg. **teruyna** RB-A 23[24].2, 23[24].3, 27[29].2, 30[32].3, 40[40].6

tewet n.m. *density*: RB-A <31[32].2>

tewhau* vn *coagulate*: pres. ind. 3 pl. **teỽhaant** RB-A <61[56].2>; pres. ind. impers. **tewheir** Pen.17 <61[56].2>

tewit pret. impers. < **tewi*** *become silent*: RB-A Add. 248v Bridging passage.7

teyrnas n.m./f. *kingdom, realm*: RB-A, 20[21].10, 21[22].2; pl. **teyrnassoed** RB-A <8[9].7>, 38[38].4

tir n.m. *land*: RB-A 12[13].8, 29[31].2

titheu affixed conjunctive pron. 2 sg. *you*: RB-A LC.7, LH.5

to n.m./f. *layer*: RB-A 12[13].13

ton n.f. *wave*: RB-A 49[45].2; pl. **tonneu** RB-A 50[46].2, Pen.17 57[52].2

ton n.m./f. *sound, tone*: Pen.17 86[81].1, 86[81].3, 86[81].4, 86[81].5 (thrice)

torri *break, shatter*: Pen.17 55[50].3

tra conj. *while*: RB-A 11[12].9, 12[13].7, Pen.17 98[92].5

tracheuyn adv. *back, backwards*: RB-A 40[40].7, 52[47].6; with infixed pron. 3 sg. m. **tra'e geuyn** RB-A 12[13].7

traeth n.m. *coast*: RB-A 9[10].8

traethu vn *speak, relate, explain*: RB-A 37[37].16; ipv. 1 pl. **traeth6n** RB-A 20[21].14, <35[36].18>; pres. ind. 1 pl. **traeth6nn** RB-A 37[37].16; pres. subj. 3 sg. **traetho** RB-A 248 Bridging passage.1

trag6res n.m. *excessive heat*: RB-A 6[6].2, 32[33].4

tragywyd adj. *eternal*: RB-A 29[31].8; **yn tragywyd** *eternally*: RB-A 2[2].5

tragywyda6l adj. *eternal*: RB-A 1[1].1; **tragywydaul** Pen.17 72[67].5

traoeruel n.m./f. *excessive cold*: RB-A 6[6].6

trasychdur n.m. *extreme dryness*: Pen.17 71[66].1

trechaf adj. *strongest, dominant*: Pen.17 67[62].2

trei n.m. *ebb*: RB-A 40[40].1

treiha pres. ind. 3 sg. < **treio*** *ebb, recede*: RB-A 40[40].2 (twice)

treulir pres. ind. impers. < **treulio*** *erode, consume*: RB-A 48[45].2

tri num. m. *three*: RB-A 30[32].7, Pen.17 87[82].3; **tri ugeint** num. *sixty*: RB-A 23[24].5, <35[36].13>, Pen.17 88[83].10; **trugein** RB-B 23[24].3, WB 23[24].3, Ph. 23[24].3; see also **trychant**

trichorna6c adj. *three-horned*: RB-A 12[13].16

trinda6t n.f. *Trinity*: RB-A Add. 248v Bridging passage.4, RB-A LC.2, LC.5

trist adj. (possibly adv. use) *sad*: RB-A LC.7

tristit n.m. *sadness, grief*: RB-A 37[37].12

troet n.m./f. *foot*: RB-A 11[12].2, 11[12].8; pl. **traet** RB-A 11[12].10, 12[13].3, 12[13].16, 12[13].17; **troetued** RB-A 12[13].19, 12[13].22

troi vn *turning, orbiting, rotating*: Pen.17 73[68].1; pres. ind. 3 sg. **try** RB-A 12[13].7, Pen.17 94[88].3; pres. ind. 3 pl. **troant** RB-A 9[10].12, Pen.17 85[80].1

tros prep. *over, across*: Pen.17 98[92].2; pers. pron. 3 sg. **trostau** Pen.17 71[66].5

truan n.m./f. *miserable*: RB-A LC.10

trugarhau vn *have mercy on*: RB-A 37[37].6

tr6m adj. *heavy* (here used as n.m.): RB-A 52[47].7; spv. **trymaf** RB-A 3[3].10, 3[3].12, RB-B 3[3].12

tr6y prep. *through*: RB-A 5[5].6, 7[7].3, 9[10].9, 30[32].1, 38[38].2, 38[38].3; **trvy** WB 38[38].2; **truy** Pen.17 81[76].4, 88[83].11, 97[91].2; **trwy** Pen.17 109[103].3, 127[121].1; with pers. pron. 3 sg. f. **tr6ydi** RB-A 5[5].5

tr6yn n.m. *nose/snout*: RB-A 11[12].11

trychant num. *three hundred*: RB-A 12[13].22, Pen.17 88[83].9, 88[83].10, RB-A Add. 248v 88[83].10

tryded num./adj. f. *third*: RB-A 2[2].5, 6[6].9, 7[7].2, 10[11].15; **trydyd** RB-B 2[2].2,

Pen.17 60[55].7, RB-A 60[55].5, RB-B 60[55].5, WB 60[55].3; **trydid** Rawl. 60[55].5

trymaf see **trʋm**

tryzor n.m./f. *treasure*: RB-A LH.3

tu n.m./f. *side*: Pen.17 60[55].2, 60[55].3, 60[55].8, 60[55].13, RB-A 60[55].5; **y tu a(c)** prep. *towards* RB-A 9[10].11, 33[34].5–6; with def. art. **(y) tu a('r)** *towards (the), facing (the)*: RB-A 9[10].12, 21[22].2, 23[24].2, 27[29].4, 33[34].5, 33[34].5–6; **e tu a'r** Pen.17 127[121].2; **ar y tu a'r** *on the side towards*: RB-A 24[25].1; **o'r tu arall** adverbial phrase *on the other side*: RB-A 6[6].4, 6[6].8; **o'r tu hʋnt (y)** adv. phrase *from beyond (the), on the other side of (the)*: RB-A 9[10].13, 32[33].3; **(y) tu hʋnt y** adverbial phrase *beyond, (on the) other side (of)*: RB-A 29[31].7, RB-B 29[31].4–5, 32[33].5–6, RB-A <35[36].8>; **y tu hvnt y** WB 29[31].4–5, 32[33].6; see also **pa du bynnac**

tʋf n.m. *waxing (of the moon)*: RB-A 74[69].18, RB A-516 74[69].4

tʋym[1] adj. *warm*: RB-A 3[3].9, WB 60[55].4; **tʋumyn** RB-B <35[36].4>; **tʋvym** WB <35[36].4>

tʋym[2] n.m. *warmth*: RB-A 6[6].6

ty n.m. *house*: Pen.17 63[58].4

tymestyl n.m./f. *tempest, storm*: WB 60[55].5, Pen.17 71[66].1; **tymhestyl** RB-A 60[55].8, WB 60[55].3; **temystyl** Pen.17 57[52].1; **temhestyl** RB-B 60[55].8; pl. **tymhestleu** RB-A 60[55].7; **tymestleu** Pen.17 60[55].9, 60[55].13; **tymhesteu** RB-A 60[55].11; **tymhestloed** RB-B 60[55].6; **tymystloed** Rawl. 60[55].8; **tymestloed** Rawl. 60[55].6

tynnʋ *drag, pull down, stretch out*: Pen.17 71[66].2; pres. ind. 3 sg. **tynn** RB-A 12[13].23, RB-A 30[32].1; pres. ind. 3 pl. **tynnant** Pen.17 <61[56].2>, RB-A <61[56].1>; pres. subj. 3 sg. **tynno** RB-A 40[40].1

tyuu vn *grow, wax (of the moon)*: RB-A 10[11].5; **tyfu** RB-A 67[62].2; pres. ind. 3 sg. **tyf** RB-A 10[11].17, 40[40].3, Pen.17 74[69].13, RB-A 74[69].11; pres. subj. 3 sg. **tyuo** RB-A 40[40].2

tywyll adj. *dark, opaque*: RB-A 37[37].3, Pen.17 74[69].11, RB-A 74[69].11, 74[69].18, RB A-516 74[69].2; **tyʋyll** RB-A 37[37].8

tywyllʋch n.m. *darkness*: RB-A 37[37].4, 74[69].5, LC.5; **tywylluc** Pen.17 71[66].4

tywynnʋ vn *to (cause to) shine/illuminate*: Pen.17 63[58].2; **tywynnu** Pen.17 63[58].4; pres. ind. 3 sg. **tywynna** RB-A <35[36].13>, Pen.17 95[89].2; impf. ind. 3 sg. **tywynnei** Pen.17 63[58].3

u/v

uch adj. *higher, above, beyond*: Pen.17 85[80].2; cpv. Pen.17 85[80].2

uchel* adj. *high*: cpv. **kyfuch** RB-A Bridging passage.4; spv. **uchaf** *highest*: RB-A 3[3].11, RB-B 3[3].14, RB-A 11[12].2, 23[24].1, 40[40].5

uchet n.m. *height*: Pen.17 74[69].10, RB-A 74[69].9

ugeint num. *twenty*: RB-A 5[5].1, 23[24].5, 29[31].4, 33[34].4, <35[36].13>; **ugein** WB 74[69].6

uffern n.f. *Hell*: RB-A <35[36].18>, 36[37].1, 37[37].12

vn num. *one*: RB-A 2[2].1, 5[5].4, 6[6].10, 7[7].1, 9[10].1; **un** RB-A 5[5].1, 12[13].4, 40[40].6, 74[69].16, RB-A Add. 248v 88[83].4; ỽn Pen.17 74[69].17, 88[83].4, 88[83].8, 94[88].1; see also **pob un/vn**

un *(the) same*: RB-A 52[47].5; **vn** RB-A 23[24].3

vndaỽt n.m./f. *unity*: RB-A LC.5

vnllygeityaỽc adj. *one-eyed*: RB-A 11[12].7

vnrym = **un** + **grym** n.m. *force*: RB-A 38[38].6

u/v/ỽ

ual prep. *like, as, similar*: RB-B <1[1].6>, RB-A 32[33].5, <35[36].2>, 37[37].8, 56[51].5; ỽal Pen.17 63[58].3, 74[69].3, 81[76].4, 93[87].3; **val** RB-A 74[69].12; with object pronoun **val y** RB-A LH.7

val hynny *thus, so*: RB-A 6[6].8, 36[37].1; ỽal henne Pen.17 87[82].1

ual (y) conj. *according as, how, like, as*: RB-A 1[1].6, RB-B <1[1].6>, RB-A 2[2].9, 37[37].2, 40[40].2, 40[40].3; ỽal (y) Pen.17 87[82].1 96[90].5; **ual (yd)** RB-A 48[45].1; **val (y)** RB-A 1[1].5, 74[69].12, LH.6; **val yd** RB-A LH.9

Uenus *Venus* (planet): Pen.17 88[83].3; (goddess) 109[103].2; ỽenus Pen.17 (planet) 86[81].2; (goddess) 109[103].4

uelly adv. *thus, likewise, so, accordingly*: RB-A 10[11].24; **velly** RB-B <1[1].4>, RB-A 37[37].3, 37[37].6; **y** ỽelly adv. *thus, likewise, so*: Pen.17 73[68].6, 98[92].4; **euelly** Pen.17 72[67].3

ỽry adj. *lofty, high*: Pen.17 85[80].3

ỽulturnus (wind): Pen.17 60[55].6; **Uulturnus** RB-A 60[55].4; **Vulturnius** RB-B 60[55].4; ỽlturus Rawl. 60[55].9

vy poss. pron. 1 sg. *my*: RB-A LC.7, LH.15; **uy** RB-A LH.10

vyg poss. pron. 1 sg. *my*: RB-A LH.16

vyn poss. pron. 1 sg. *my*: RB-A 2[2].11

w/ỽ

weithon adv. *now, henceforth*: RB-A 20[21].14, 29[31].8, 32[33].10, <35[36].18>; **weithyon** Pen.17 57[52].4; **weitheon** Pen.17 88[83].9, 88[83].12; **withyon** RB-A 37[37].16

withyon see **weithon**

ỽrth prep. *close to, by the side of, opposite of, facing, because of*: RB-A 1[1].3, 52[47].8, RB-B 60[55].3, Rawl. 60[55].3; **urth** Pen.17 56[51].6, 74[69].2, 74[69].5, 87[82].4; with pers. pron. 3 pl. ỽrthunt RB-A 37[37].6; **wrth hynny** conjunctive adverbial phrase *therefore, thereupon, accordingly, because of that*: WB 74[69].3; ỽrth hynny RB-A 5[5].7, 49[45].1, 52[47].8, 74[69].4, RB-B 74[69].2;

vrth hynny WB 74[69].1; **urth henne** Pen.17 58[53].2, 74[69].5

ỽy affixed pron 3 pl. *they*: RB-A LH.18, Add. 248v Bridging passage.6

wy n.m. *egg*: RB-A 1[1].2, 1[1].3; **ỽy** RB-B <1[1].2>

wybyr n.f. *sky, cloud*: RB-A 1[1].6 (twice), Pen.17 57[52].2, 60[55].4, 60[55].14; **ỽybyr** RB-A 37[37].4, WB 60[55].8, Rawl. 60[55].1, 60[55].3, 60[55].10, <61[56].2>; **wyber** Pen.17 60[55].2, 60[55].6; **wybren** Pen.17 63[58].2 (twice), 74[69].6, 145[138].1; pl. **wybrenneu** RB-A 60[55].1

ỽyth num. *eight*: RB-A 11[12].2, RB-A Add. 248v 88[83].2, 88[83].6, 88[83].7; **wyth** Pen.17 74[69].16, RB-A 74[69].15, Pen.17 88[83].7; **vyth** Pen.17 88[83].3

wythuet num. *eighth*: RB-A, 10[11].16, 11[12].6, LC.4

wyneb n.m. *face* RB-A 74[69].12

wyneb yn wyneb *face to face*: RB-A 33[34].5

ỽynt pers. pron. 3 pl.: RB-A 11[12].8; **wynt** RB-A <61[56].3>, Pen.17 81[76].4

ỽynteu affixed pers. pron. 3 pl.: RB-A 37[37].6; **wynteu** Pen.17 56[51].5, 73[68].2, 96[90].2 (twice), 109[103].2

y

y[1] def. art. *the*: RB-A 1[1].1, 1[1].3 (twice), 1[1].4 (thrice); **i** RB-B 23[24].4; **e** Pen.17 55[50].1, 56[51].1, 56[51].3, 56[51].5 (twice)

y[2] poss. pron. 3 sg.m.: RB-A 1[1].2, 2[2].7, 6[6].5, 12[13].9, 12[13].14; **e** Pen.17 60[55].6, 74[69].19; poss. pron. 3 sg. f. RB-A 5[5].3, 5[5].4, 8[8].1, 10[11].1, 12[13].18; **e** RB-A 30[32].5, Pen.17 93[87].2, 109[103].2; poss. pron. 3 pl. **y** RB-A 32[33].8

y[3] inf. obj. pron. (used with **pan**): RB-A 40[40].1

y[4] prep. *to, for*: 8[8].5, <8[9].7>, 9[10].9, 9[10].10, 10[11].22; (*in order*) *to*: RB-A 5[5].6, 52[47].4; **e** Pen.17 71[66].5; with suffixed personal pron. 1 sg. **ymi** RB-A LH.9; 1 pl. **in** RB-A 37[37].16; 2 sg. **itti** RB-A LC.9; **itt** RB-A LH.7; **ytti** RB-A LH.16; 3 sg. m. **idaỽ** *to it*: RB-A 12[13].4, 12[13].6 (twice), 12[13].13, 12[13].17, 12[13].20; **idav** RB-A 12[13].2; **idau** Pen.17 60[55].2, 60[55].8, 98[92].4, 128[122].1; 3 sg. f. **ydi** WB 23[24].5; **idi** RB-B 23[24].3, Ph. 23[24].5, RB-A 42[42].3, 74[69].4, 74[69].10; 1 pl. **yni** RB-A 6[6].10; 3 pl. **udunt** *to them*: RB-A 10[11].20, 11[12].2, 11[12].3, 11[12].8, 11[12].9; with def. art. **y'r** RB-A 2[2].2, RB-B 2[2].1, RB-A 3[3].8, 3[3].9, 3[3].13 (twice); **e'r** Pen.17 56[51].7, 57[52].4, 67[62].1, 74[69].2

y[5] preverbal part.: RB-A 1[1].6, RB-B <1[1].3>, RB-A 2[2].2, 2[2].4, 2[2].7 (twice); **e** Pen.17 74[69].7; **yd** RB-A 2[2].1, 2[2].2, <8[9].7>, 9[10].7, 10[11].2, 10[11].3; **yt** Pen.17 109[103].1; **et** preverbal part.: Pen.17 73[68].4; with obj. pron. 3 sg. m. **y** RB-B 3[3].10, 3[3].11, RB-A 12[13].24, RB-B 32[33].12, WB 32[33].12; with obj. prob. 3 sg. f. **y** RB-B 3[3].8

y[6] rel. part.: RB-B 3[3].1, RB-A 11[12].14, 20[21].6, <31[32].4>, 40[40].4

y[7] part. used with vn: RB-A 9[10].10

y ar prep. *on, upon*: Pen.17 60[55].15, 71[66].4, RB-A Add. 248v Bridging passage.4, Pen.17 145[138].1, 146[139].1; with pers. pron. 3 pl. **y arnu** *on them, upon them*: RB-A 29[31].5

y danaỽ prep. *under* with 3 sg. m. pron.: RB-A 81[76].2

y gan see gan

y gilyd see kilyd

y mae see bot

y myỽn see myỽn

y'r see y⁴

y rei see rei

y rỽng prep. *between, among*: RB-A 6[6].6, 27[29].5; yr rỽng RB-A 10[11].10; y rwg WB 32[33].4; with pers. pron. 1 pl. y rom *between us*: Pen.17 94[88].3; *for our sake* RB-A LC.11

y tu hỽnt y see tu

y velly see uelly

y ỽrth prep. *from*: RB-A 2[2].9, 2[2].10, RB-B 32[33].10, RB-A 74[69].11, RB A-516 74[69].2; y urth Pen.17 74[69].11 (twice); with poss. pron. 3 sg. m. y ỽrthav RB-A 40[40].1; y ỽrthaỽ RB-A 40[40].6; with poss. pron. 3 sg. f. y ỽrthi RB-A 74[69].10, RB A-516 74[69].1

y urthuynep see gwrthỽyneb

ya n.m. *ice, frost*: Pen.17 60[55].4

yaen n.f. *sheet or cake of ice*: Pen.17 93[87].3

yaennỽ vn *turn into ice*: Pen.17 65[60].1

yaỽn n.m. *rightness, truth*: Pen.17 92[86].3

ychen n. pl. < ych* n.m. *ox* RB-A 12[13].16

yd see y⁴

yuet *to drink, drinking*: RB-B 32[33].3, WB 32[33].3, Pen.17 71[66].3; pres. ind. 3 pl. yfant RB-A 10[11].25

yg see yn³

ygkylch *around* RB-A 1[1].3, 1[1].4 (twice), 1[1].5, 1[1].6 (twice)

ym blaen prep. *in front of*: RB-A <35[36].11>

ymadraỽd n.m./f. *speech, language*: RB-A 11[12].4, 12[13].5

Ymago Mundi (Lat.) *Image of the World*: RB-B Intro.1, WB 81[76].3

yman adv. *here*: RB-A 3[3].14

ymberiglaỽ vn *imperil oneself*: RB-A LC.5

ymborth n.m./f. *food, nourishment*: RB-A 11[12].13, 12[13].15

ymborthant pres. ind. 3 sg. < ymborth* *feed, nourish (oneself)*: RB-A 10[11].24

ymchuelut vn *to turn, return, change into*: Pen.17 67[62].3; pres. ind. 3 sg. ymchoelut RB-A 52[47].7, 67[62].2, RB A-516 67[62].2; pres. ind. impers. ymchoelir RB-A 3[3].4; pret. 3 sg. ymchoel RB-A 48[45].3

ymdaant pres. ind. 3 pl. < ymdaith* *walk*: RB-B 3[3].17

ymdianc vn *flee*: RB-A 42[42].2

ymdodant pres. ind. 3 sg. < ymdoddi* *digest/dissolve*: RB-A LH.17

ymdywynyco press. subj. 3 sg. < **ymdywynygu*** *shine, appear*: Pen.17 57[52].1

ymddangos* *appear, show oneself*: pres. ind. 3 sg. **ymdengys** Pen.17 63[58].3; **ymdangossant** pres. ind. 3 pl. Pen.17 95[89].1; pres. subj. 3 pl. **ymdangossont** Pen.17 58[53].5

ymgyffelybu vn *be similar, resemble*: Pen.17 87[82].5

ymgyuodi vn *rise, get up*: Pen.17 57[52].2

ymgymysgu* *to mix*: pres. ind. 3 pl. **ymgymysgant** RB-A 3[3].6; pres. subj. 3 sg. **ymgymysco** RB-A 48[45].1

ymgyrhaedei impf. subj. 3 sg. < **ymgyrhaeddu*** *approach, grasp*: RB-A 6[6].6

ymlad n.m. *battle, strife*: RB-A 10[11].16, 12[13].8

ymladd* *(to do) battle, to fight*: pres. ind. 3 sg. **ymlad** RB-A 12[13].7; pres. ind. 3 pl. **ymladant** RB-A 10[11].19, 12[13].10

ym perued see **perued**

ymplith prep. *amongst, (together) with*: RB-A 10[11].17, Pen.17 98[92].3, 109[103].4, 128[122].2, 140[134].2

ymrⱱymant pres. ind. 3 pl. < **ymrwymo*** *to bind oneself*: RB-A 3[3].6

yn[1] adv. part.: RB-A 1[1].1, 2[2].5, 2[2].8, 3[3].7, 9[10].8; **en** Pen.17 56[51].6, 97[91].2

yn[2] particle used with vn: RB-A 12[13].15, 32[33].5, 34[35].1, <35[36].9>, RB-B 34[35].2; **en** Pen.17 58[53].4, 71[66].2 (twice), 73[68].1, 109[103].1

yn[3], **yg**, **ym**, **y[n]**, **y[m]** prep. *in, into*: RB-A 1[1].5, RB-B <1[1].5> (thrice), <1[1].6> (thrice), RB-A 2[2].2, 2[2].6; **en** Pen.17 56[51].1, 56[51].2, 56[51].5, 58[53].3, 60[55].4; with suffixed personal pron. 3 sg. m. **yndaⱱ** *in it*: RB-A 2[2].3, 34[35].2, RB-B 34[35].2, RB-A 49[45].1, LH.21; **yndav** WB 34[35].2; **endau** Pen.17 55[50].1, 58[53].3 (twice), 72[67].1, 98[92].1, 143[136].1; **endaⱱ** Pen.17 73[68].1; with suffixed pers. pron. 3 sg. f. **yndi** *in it*: RB-A 5[5].7, 10[11].5, WB 23[24].3, Phil. 23[24].3, RB-B 29[31].4; **endi** Pen.17 74[69].6; pl. **yndunt** RB-A 11[12].1, 48[45].3; **endunt** Pen.17 56[51].6

yn[4] predicative part.: RB-A 1[1].2, 2[2].6, 10[11].17, 11[12].2, 11[12].3, 11[12].5; **en** Pen.17 56[51].3, 58[53].1, 65[60].2, 67[62].3 (twice)

yn erbyn prep. *towards, against* (prep. **yn** *in* + prep. **erbyn** *towards*): RB-A 9[10].4, 27[29].1, 74[69].14, 81[76].1, 81[76].1-2; **en erbyn** Pen.17 73[68].1, 73[68].4, 73[68].5, 73[68].6; **erbyn yn erbyn** *face to face*: RB-A Add. 248v Bridging passage.3, Pen.17 97[91].1-2

yn eu kylch see **ygkylch**

yn lle *instead of, in lieu of*: RB-A 11[12].4, 11[12].11, 12[13].4; **en lle** Pen.17 98[92].3

yn wastat adv. *continually, always*: RB-A 10[11].7, 20[21].8, 29[31].6 (twice), RB-B 29[31].4, WB 29[31].4

yn y gilyd *into the other*: RB-A 3[3].3

yna adv. *then, subsequently, at that time*: RB-A 3[3].1, 52[47].7, Pen.17 74[69].12, RB-A 74[69].11, RB A-516 74[69].2; **ena** Pen.17 63[58].3

ynheu affixed pron. 1 sg.: RB-A LC.7

yno adv. *there*: RB-A <8[9].1>, <8[9].3>, 10[11].7, 10[11].8, 10[11].18, 10[11].21; **eno** Pen.17 <61[56].2>, 92[86].2 (twice), 146[139].1, 146[139].2

ynteu affixed pron. 3 sg. m.: RB-A 60[55].8, LH.10; **enteu** Pen.17 56[51].2 60[55].2, 60[55].5, 60[55].6, 60[55].8, 60[55].13; see also **enteu**[1]

ynys n.f. *island*: RB-A 10[11].4, 10[11].5, 10[11].6, 20[21].13, 23[24].6; pl. **ynyssed** RB-A 10[11].14, 29[31].1, 29[31].3, RB-B 29[31].1, RB-A 32[33].10; **ynyssoed** WB 29[31].1, RB-A 33[34].1

yr[1] def. art. *the*: RB-A 1[1].3, 1[1].5, 1[1].6 (thrice), RB-B <1[1].6>; **er** Pen.17 56[51].2 (twice), 56[51].3, 60[55].4, 60[55].14, 63[58].1

yr[2] prep. (used with eq. adj., *GMW*, §45) *however*: RB-A 11[12].13

yr[3] prep. *during, because of, despite*: RB-A 48[45].1, 49[45].1, Pen.17 55[50].1, RB-A Add. 248v Bridging passage.6

yr aⱱr honn *presently, now*: RB-A 2[2].11, <35[36].10>

yr hⱱnn def. art. + dem., used as rel. pron.: RB-B <1[1].2>, RB-A LH.1, LH.22; **yr hwnn** RB-A 9[10].13; **er hwn** Pen.17 96[90].3; **yr honn** RB-A 74[69].2; **er honn** Pen.17 127[121].1

yscaⱱn adj. *light* (not heavy), *swift*: RB-A 52[47].2, 52[47].7; spv. **yscaⱱnaf** *lightest*: RB-A 3[3].10; **ysgaⱱnaf** RB-A 3[3].13, RB-B 3[3].13

yscriuennu vn *writing, to write*: RB-A <35[36].9>; pres. ind. impers. **yscriuennir** RB-A 2[2].1; **ysgriuennir** RB-A 52[47].3; **yscriuenir** Pen.17 146[139].3

yscruthur n.f. *scripture*: RB-A LH.5

yscythredic verbal adj. *decorated*: Pen.17 93[87].2

ysgaelussaⱱ vn *ignore, neglect, forsake*: RB-A LH.17

ysgythreis pret. 1 sg. < **ysgythru*** *to depict*: RB-A LH.6

yspeit n.m./f. *period, interval*: Pen.17 81[76].4, RB-A 81[76].2

Yspryt Glan *Holy Spirit*: RB-A LC.3

ysprytoed pl. < **ysbryd*** n.m. *spirit, soul*: RB-A 37[37].11

ysso pres. subj. 3 sg. < **ysu** *to eat*: RB-B <35[36].3>, WB <35[36].3>

yssyd see **bot**

yt see **yd**

ytti see **y**

yttⱱyt see **bot**

yv/yⱱ see **bot**

Z

Zephirus *Zephyr* (wind): Pen.17 60[55].11, RB-B 60[55].9, WB 60[55].6; **Zepherus** RB-A 60[55].9; **Sephirus** Rawl. 60[55].9

zodiacus *Zodiac*: Pen.17 74[69].16, RB-A Add. 248v Bridging passage.1, Pen.17 97[91].2, 127[121].2; **zodiatus** RB-A 81[76].2

INDICES

For the indices below all instances and spelling variants are included. Where possible modern identifications or modernisations of the name are given. Where this is not possible, the equivalent Latin form is provided.

INDEX OF PLACE NAMES

Corsica *Corsica* WB <35[36].5>; see also Sirme and Cursita

Creta *Crete* RB-A 33[34].2

Cursita *Corsica* RB-B <35[36].5>

Cyprys *Cyprus* RB-A 33[34].2

Cyrene *Cyrene* RB-A 30[32].6

Danubius see Auon Danubi

Denmarc *Denmark* RB-A 24[25].2

Denos *Delos* RB-A 33[34].7

Dinas Cerocemus *City of Parethonium* RB-A 30[32].3

Dinas Cyrene *Cyrene* (city) RB-A <35[36].12>

Dinas Tharsus *Tarsus* (city) RB-A 20[21].11

Dinas Yponensis *civitas Yppone* RB-A <31[32].4>

Edos *Cohos?* RB-A 33[34].4

Eifft *Egypt* RB-A 9[10].9, 33[34].6

Ergete (island) *Argere* RB-A 10[11].6

Esperia *Hesperia* RB-B 28[30].1, **Ysteria** WB 28[30].1

Ethiopia *Ethiopia* RB-A 32[33].1 (twice), 32[33].3, RB-B 32[33].1, WB 32[33].1, RB-B 32[33].6, WB 32[33].6, RB-A <35[36].11>

Euffrates (river) *Euphrates* RB-A 9[10].11

Europa *Europe* RB-A 7[7].2, 7[7].3, 20[21].15, WB 21[22].1, Ph. 21[22].1, RB-A 33[34].4, <35[36].10>, **Evroppa** RB-B 21[22].1, **Europam** RB-A 29[31].8

Ffreinc *France* RB-A 23[24].7, WB 23[24].4, Ph. 23[24].4, RB-A 27[29].2, 27[29].3, RB-B 27[29].2, WB 27[29].2

Ffreinc Liỽn *Ludgunensis Gallia* RB-A 27[29].4

Ffrigia Uỽyaf *Phrygia Maior/Greater Phrygia* RB-A 20[21].1

Ffrigya *Phrygia* RB-A 20[21].4

Ffyson (river) *Physon* RB-A 9[10].1

Gades *Cadiz* RB-A 32[33].6, **Kagades** RB-B 32[33].9, **Ccaer Gades** WB 32[33].9

Galathia *Galatia* RB-A 20[21].2

Gallia *Gaul* RB-A 27[29].1, **Agallica, Bellica**

Gallia Belgica RB-B 27[29].1, **Gallia Bellica** WB 27[29].1

Gallicia *Galicia* RB-A 28[30].2, RB-B 28[30].2, WB 28[30].2

Ganges (river) *Ganges* RB-A 9[10].2, 11[12].12, **Auon Gange** 12[13].21

Germania *Germania/Germany* RB-B 23[24].1

Germania Issaf *Germania Inferior/Lower Germany* RB-A 24[25].1

Germania Uchaf *Germania Superior/Upper Germany* RB-A 23[24].1

Germania Vawr *Gemania Maior* WB 23[24].1

Getulia *Getulia* RB-A <31[32].3>

Gỽasgỽyn *Gascony* RB-A 27[29].5

Gỽlat y Blammonyeit *Ethiopia* RB-A 9[10].9

Gyon (river) *Geon/Gihon* (= Nile) RB-A 9[10].5

Hellesponto *Hellespont* (= Dardanelles) RB-A 33[34].3

Hybernia *Ireland* WB 29[31].2

Iberia *Iberia* RB-B 28[30].1, **Yberia** WB 28[30].1

India *India* RB-A 9[10].4, 10[11].1, 10[11].3, 10[11].9, 10[11].13, Pen.17 60[55].14, RB-A 60[55].12, RB-B 60[55].11, Rawl. 60[55].11, **Yndea** WB 60[55].8

Indus (river) *Indus* RB-A 10[11].1, 30[32].1

Iwerdon *Ireland* RB-A 29[31].2, RB-B 29[31].2; see also **Hybernia**

Leptis Vaỽr *Leptis Magna* RB-A 30[32].8

Libia *Libya* RB-A 30[32].2

Licaonia *Lycaonia* RB-A 20[21].6

Lidia *Lydia* RB-A 20[21].7

Liger *Loire* RB-A 27[29].5

Lithia *Lycia* RB-A 20[21].12

Lloegyr *England* RB-B 29[31].2; see also **Anglia**

Llychlyn *Norway* RB-A 24[25].2

Lucitania *Lusitania/Histapia Lucitana* RB-B 28[30].2, **Llucitammia** WB 28[30].2

INDEX OF PEOPLES

INDEX OF PERSONAL NAMES

BIBLIOGRAPHY

Manuscripts

MSS of Delw y Byd

Aberystwyth, NLW, MS 5267B
Aberystwyth, NLW, MS Peniarth 5 'White Book of Rhydderch'
Aberystwyth, NLW, MS Peniarth 17
Cardiff, Central Library, 2.83
Oxford, Bodleian Library, Rawlinson B 467
Oxford, Jesus College, MS 111 'Red Book of Hergest'
Philadelphia, Library Company of Philadelphia, MS 8680.O

MSS of Imago Mundi

Brussels, Bibliothèque Royale, MS 2419–31
Brussels, Bibliothèque Royale, MS 10862–5
Cambridge, Corpus Christi College MS 66
Exeter, Cathedral Library, MS 3514
London, British Library, Additional MS 38665
London, British Library, Cotton Cleopatra B IV
London, British Library, Royal 13 A xxi
Oxford, Bodleian Library, Rawlinson B 484
Paris, Bibliothèque de l'Arsenal, Arsenal 93(B)
Paris, Bibliothèque Mazarine, MS 708
Paris, BnF, MS lat. 11130
Paris, BnF, MS lat. 15009

Other MSS

Aberystwyth, NLW, MS Peniarth 14
Cardiff, Central Library, MS 2.81 'Book of Aneirin'
Cardiff, Central Library, MS 3.242 (Hafod 16)
Oxford, Jesus College, MS 20

Primary Sources

AUGUSTINE OF HIPPO, *De Genesi ad litteram libri duodecim. La génèse au sens littéral en douze livres*, ed. by P. Agaësse and A. Solignac, Bibliothèque Augustinienne 48–49, 2 vols (Paris: Desclée De Brouwer, 1972; repr. Turnhout: Brepols, 2001)

—— *City of God: XVI–XVIII.35*, trans. by Eva Matthews Sanford and William McAllen Green, Loeb Classical Library 415 (Cambridge, MA: Harvard University Press, 1965)

BARRON, W. R. J., and GLYN S. BURGESS, eds, *The Voyage of Saint Brendan:*

Representative Versions of the Legend in English Translation (Exeter: University of Exeter Press, 2002)

BROMWICH, RACHEL, ed. and trans., *Trioedd Ynys Prydein: The Triads of the Island of Britain*, 3rd edn (Cardiff: University of Wales Press, 2006)

BROMWICH, RACHEL, and D. SIMON EVANS, eds, *Culhwch ac Olwen: An Edition and Study of the Oldest Arthurian Tale* (Cardiff: University of Wales Press, 1992)

CLANCY, THOMAS OWEN, trans., *The Triumph Tree: Scotland's Earliest Poetry AD 550–1350* (Edinburgh: Canongate Books, 1998)

DANIEL, R. IESTYN, ed., *Gwaith Dafydd y Coed a Beirdd Eraill o Llyfr Coch Hergest*, Cyfres Beirdd yr Uchelwyr (Aberystwyth: University of Wales Centre for Advanced Welsh and Celtic Studies, 2002)

DUMVILLE, DAVID N., ed. and trans., *Brenhinoedd y Saeson = 'The Kings of the English', A.D. 682–954: texts P, R, S in parallel* (Aberdeen: University of Aberdeen, 2005)

EVANS, D. SIMON, ed., *Historia Gruffud vab Kenan* (Cardiff: University of Wales Press, 1977)

—— *A Medieval Prince of Wales: The Life of Gruffudd ap Cynan* (Lampeter: Llanerch, 1990)

FALILEYEV, A., ed., *Welsh Walter of Henley*, Medieval and Modern Welsh Series 12 (Dublin: Dublin Institute for Advanced Studies, 2006)

GOSSOUIN DE METZ, *L'Image du Monde*, ed. by O. H. Prior (Lausanne and Paris: Payot, 1913)

HAYCOCK, MARGED, ed., *Blodeugerdd Barddas o Ganu Crefyddol Cynnar* (Abertawe: Cyhoeddiadau Barddas, 1994)

—— ed. and trans., *Legendary Poems from the Book of Taliesin*, 2nd edn (Aberystwyth: CMCS, 2015)

HERODOTUS, *Herodoti Historiae. Libri I–IV*, ed. by N. G. Wilson (Oxford: Oxford University Press, 2015, first published 1902, rev. edn 1927)

HONORIUS AUGUSTODUNENSIS, *De imagine mundi libri tres*, ed. by Jean-Paul Migne, Patrologiae Cursus Completus, Series Latina 172 (Paris: J.-P. Migne, 1895)

—— *Imago mundi*, ed. by V. I. J. Flint, Archives d'histoire doctrinale et littéraire du Moyen Âge 49 (Paris: Librairie Philosophique J. Vrin, 1983)

ISAAC, G. R., and S. RODWAY, 'Peniarth 17', *Rhyddiaith Gymraeg o Lawysgrifau'r 13eg Ganrif: Testyn Cyflawn* (2002) <http://cadair.aber.ac.uk/dspace/handle/2160/5829>

—— 'Peniarth 17', in G. R. Isaac, Simon Rodway, Silva Nurmio, Kit Kapphahn, and Patrick Sims-Williams, *Rhyddiaith Gymraeg o Lawysgrifau'r 13eg Ganrif: Fersiwn 2.0* (2010) <https://cadair.aber.ac.uk/dspace/handle/2160/11163>

JEROME, 'Epistola LVII. Ad Pammachium; De optimo genere interpretandi', in *Sancti Hieronymi Stridonensis Presbyteri Opera Omnia I*, ed. by J.-P. Migne, Patrologiae cursus completus. Series latina XX (Paris: J.-P. Migne, 1845), cols 568–79

JONES, ELIN M., and NERYS ANN JONES, eds, *Gwaith Llywarch ap Llywelyn 'Prydydd y Moch'* (Cardiff: University of Wales Press, 1991)

JONES, GWENAN, ed., '*Gwyrthyeu y Wynvyvydedic Veir*', *BBCS*, 9 (1937–1939), 114–18, 334–41; 10 (1939–1941), 21–23

JONES, J. MORRIS, and JOHN RHŶS, eds, *The Elucidarium and Other Tracts in*

Welsh, from Llyvr Agkyr Llandewivrevi A.D. 1346 (Jesus College MS. 119) (Oxford: Clarendon Press, 1894)

JONES, THOMAS, ed. and trans., *Brut y Tywysogyon or The Chronicle of the Princes: Red Book of Hergest Version*, History and Law Series 16 (Cardiff: University of Wales Press, 1955)

LEWIS, HENRY, 'Y Diarhebion ym Mheniarth 17', *BBCS*, 4 (1927–1929), 1–17

—— ed., *Brut Dingestow* (Cardiff: University of Wales Press, 1942)

—— and P. DIVERRES, eds, *Delw y Byd (Imago Mundi)* (Cardiff: University of Wales Press, 1928)

LUFT, DIANA, PETER WYNN THOMAS, and D. MARK SMITH, eds, *Rhyddiaith Gymraeg 1300–1425* (2013) <http://www.rhyddiaithganoloesol.caerdydd.ac.uk>

MANZALAOUI, M. A., ed., *Secretum Secretorum: Nine English Versions*, EETS Original Series 276 (Oxford: Oxford University Press, 1977)

MOSELEY, C., *Mandeville's Travels* (Harmondsworth: Penguin, 2005)

PLINY, *Natural History: Books 3–7*, ed. by J. Henderson, trans. by H. Rackham, Loeb Classical Library 352 (Cambridge, MA, and London: Loeb, 1942)

POPPE, ERICH, and R. RECK, eds *Selections from Ystorya Bown o Hamtwn* (Cardiff: University of Wales Press, 2009)

ROBERTS, BRYNLEY F., *Astudiaeth destuonol o'r tri chyfieithiad Cymraeg cynharaf o 'Historia Regum Britanniae' Sieffre o Fynwy, ynghyd ag 'agraffiad' beirniadol o destun Peniarth 44* (unpublished PhD thesis, University of Wales, 1969)

—— ed., *Brut y Brenhinedd. Llanstephan MS. 1 Version*, Medieval and Modern Welsh Series V (Dublin: Dublin Institute for Advanced Studies, 1971)

—— ed., *Cyfranc Lludd a Llefelys*, Medieval and Modern Welsh Series VII (Dublin: Dublin Institute for Advanced Studies, 1975)

ROBERTS, RICHARD GLYN, ed., *Diarhebion Llyfr Coch Hergest* (Aberystwyth: CMCS, 2013)

RUSSELL, PAUL, ed., *Vita Griffini Filii Conani: The Medieval Latin Life of Gruffudd ap Cynan* (Cardiff: University of Wales Press, 2005)

SHAFF, PHILLIP, ed., *Nicene and Post-Nicene Fathers: First Series*, Volume IV, *St. Augustine* (New York: Cosimo, 2007)

SHARPE, RICHARD, and JOHN REUBEN DAVIES, eds and trans., 'Rhygyfarch's *Life* of St David', in *St David of Wales: Cult, Church and Nation*, ed. by J. Wynn Evans and Jonathan M. Wooding (Woodbridge: Boydell Press, 2007), pp. 107–55

WILLIAMS, PATRICIA, ed., *Historical Texts from Medieval Wales*, MHRA Library of Medieval Welsh Literature (London: MHRA, 2012)

WILLIAMS, STEPHEN J., ed., *Ystorya de Carolo Magno o Llyfr Coch Hergest* (Cardiff: University of Wales Press, 1968)

WILMANS, ROGER, ed., 'Ex Honorii Augustodunensis summa totius et imagine mundi', Monumenta Germaniae Historica Scriptores XII (Hannover, 1852), pp. 125–34

Secondary Sources

ALEXANDER, PHILIP, 'Jerusalem as the Omphalos of the World: On the History of a Geographical Concept', in *Jerusalem: Its Sanctity and Centrality to Judaism, Christianity, and Islam*, ed. by Lee I. Levine (New York: Continuum, 1999), pp. 104–19

ARN, MARY-JO, 'On Punctuating Medieval Texts', *Text*, 7 (1994), 161–74

AUSTIN, M. M., ed., *The Hellenistic World from Alexander to the Roman Conquest: A Selection of Ancient Sources in Translation* (Cambridge: Cambridge University Press, 1981)

BALL, MARTIN J., and NICOLE MÜLLER, *Mutation in Welsh* (London and New York: Routledge, 1992)

BARBER, RICHARD, and ANNE RICHES, *A Dictionary of Fabulous Beasts* (Woodbridge: Boydell, 1971)

BAUERREISS, R., 'Zur Herkunft des Honorius Augustodunensis', *Studien und Mitteilungen zur Geschichte des Benediktiner-Ordens*, 53 (1935), 28–36

BERTOLA, F., 'The Milky Way Through the Ages: An Iconographic Journey', in *Cosmology Across Cultures ASP Conference Series 409: Proceedings of the Conference Held 8–12 September 2009 at Parque de las Ciencias, Granada, Spain*, ed. by J. Rubiño-Martín et al. (San Francisco, CA: Astronomical Society of the Pacific, 2009), pp. 237–41

BEYER DE RYKE, BENOÎT, 'Le miroir du monde: un parcours dans l'encyclopédisme médiéval', *Revue belge de philologie et d'histoire*, 81 (2003), 1243–75

BLAISE, ALBERT, *Manuel du latin chrétien* (Strasbourg: Le Latin chrétien, 1955)

BLUME, CLEMENS, 'Hymnodia Hiberno-celtica', *Analecta Hymnica Medii Aevi*, 51 (1908) 257–365

BOULOUX, NATHALIE, 'Les îles dans les descriptions géographiques et les cartes du Moyen Âge', *Médiévales*, 47 (2004), 47–62

BRACKE, WOUTER, 'Pomponius Mela, Étienne de Byzance, Honorius d'Autun et le MS. de Bruxelles, BR 2419–31', *Revue belge de philologie et d'histoire*, 81 (2003), 1075–81

BREATNACH, PÁDRAIG A., 'The Origins of the Irish Monastic Tradition at Ratisbon (Regensburg)', *Celtica*, 13 (1980), 58–77

VON DEN BRINCKEN, ANNA DOROTHEE, 'Imago Mundi. Marginalien zum „Weltbild" des Honorius Augustodunensis (insbes. Imago Mundi I,1 und 5–7)', in *Scientia und ars im Hoch und Spätmittelalter, Festschrift für Albert Zimmermann zum 65. Geburtstag*, ed. by Ingrid Craemer-Ruegenberg and Andreas Speed, vol. 2 (Berlin and New York: De Gruyter, 1994), pp. 819–28

British Library <http://www.bl.uk> [accessed 30 March 2016]

BRUMBLE, H. DAVID, *Classical Myths and Legends in the Middle Ages and Renaissance: A Dictionary of Allegorical Meanings* (London: Greenwood, 1998)

BULLOUGH, DONALD A., *Carolingian Renewal: Sources and Heritage* (Manchester: Manchester University Press, 1991)

BURGESS, GLYN S., and CLARA STRIJBOSCH, *The Legend of St Brendan: A Critical Bibliography* (Dublin: Royal Irish Academy, 2000)

CALAME, CLAUDE, 'Structuralism and Semiotics. Narrating the Foundation of a City: The Symbolic Birth of Cyrene', in *Approaches to Greek Myth*, ed. by L. Edmunds (Baltimore and London: Johns Hopkins University Press, 1990), pp. 275–341

CALLE MARTÍN, JAVIER, 'Punctuation Practice in a 15th-century Arithmetical Treatise (MS. Bodley 790)', *Neuphilologische Mitteilungen*, 105 (204), 407–22

CAMPBELL, MARY BAINE, 'Asia, Africa, Abyssinia: Writing the Land of Prester John', in *Travel Writing, Form, and Empire: The Poetics and Politics of Mobility*, ed. by Julia Kuehn and Paul Smethurst (New York and London: Routledge, 2012), pp. 21–37

CHAPMAN, ADAM, *Welsh Soldiers in the Later Middle Ages* (Woodbridge: Boydell Press, 2015)

CHARLES-EDWARDS, GIFFORD, 'The Scribes of the Red Book of Hergest', *National Library of Wales Journal*, 21 (1980), 246–56

CHARLES-EDWARDS, THOMAS, '"Mi a dynghaf dynghed" and Related Problems', in *Hispano-Gallo-Brittonica: Essays in Honour of Professor D. Ellis Evans on the Occasion of his Sixty-Fifth Birthday*, ed. by Joseph F. Eska, R. Geraint Gruffydd, and Nicolas Jacobs (Cardiff: University of Wales Press, 1995), pp. 1–15

CHARLES-EDWARDS, THOMAS, and PAUL RUSSELL, 'The Hendregadredd Manuscript and the Orthography and Phonology of Welsh in the Early Fourteenth Century', *National Library of Wales Journal*, 28 (1993–1994), 419–62

CLARK, JAMES G., FRANK T. COULSON, and KATHRYN L. McKINLEY, eds, *Ovid in the Middle Ages* (Cambridge: Cambridge University Press, 2011)

CLARKE, MICHAEL, 'The Lore of the Monstrous Races in the Developing Text of the Irish *Sex Aetates Mundi*', *CMCS*, 63 (2012), 15–49

CLEMENS, RAYMOND, and TIMOTHY GRAHAM, *Introduction to Manuscript Studies* (Ithaca, NY: Cornell University Press, 2007)

CONNOCHIE-BOURGNE, CHANTAL, '*Comment li element sont assis*. L'image de l'œuf cosmique dans quelques encyclopédies en langue vulgaire du XIIIe siècle', in *Les quatre éléments dans la culture médiévale*, ed. by Danielle Buschinger and A. Crépin (Göppingen: Kümmerle, 1983), pp. 37–48

CRICK, JULIA, 'The Power and the Glory: Conquest and Cosmology in Edwardian Wales (Exeter Cathedral Library, MS. 3514)', in *Textual Studies: Cultural Texts*, ed. by Elain Treharne and Orietta da Rold (Cambridge: D. S. Brewer, 2010), pp. 21–42.

CROUSE, ROBERT DARWIN, 'Honorius Augustodunensis: Disciple of Anselm?', *Analecta Anselmiana*, 4 (1975), 131–39

DAMPIER, W. C., *A History of Science and its Relations with Philosophy and Religion*, 4th edn (Cambridge: Cambridge University Press, 1971)

DARIAN, S. G., 'The Ganges and the Rivers of Eden', *Asiatische Studien / Etudes Asiatiques*, 31 (1977), 42–54

DRONKE, PETER, *Fabula: Explorations into the Uses of Myth in Medieval Platonism*, Mittellateinische Studien und Texte IX (Leiden and Cologne: E. J. Brill, 1974)

EISLER, WILLIAM, *The Furthest Shore: Images of Terra Australis from the Middle Ages to Captain Cook* (Cambridge: Cambridge University Press, 1995)

EVANS, D. SIMON, *A Grammar of Middle Welsh* (Dublin: Dublin Institute for Advanced Studies, 1964)

——*Medieval Religious Literature*, Writers of Wales (Cardiff: University of Wales Press, 1986)

EVANS, J. GWENOGVRYN, *Report on Manuscripts in the Welsh Language*, 2 vols (London: for H.M.S.O. by Eyre and Spottiswoode, 1898–1905)

FALILEYEV, ALEXANDER, '*Delw y Byd* Revisited', *Studia Celtica*, 44 (2010), 71–78

FARMER, DAVID, *The Oxford Dictionary of Saints*, 5th edn, revised (Oxford: Oxford University Press, 2011)

FEJFER, JANE, 'Cyrene and the Cyrenaica', in *The Encyclopedia of Ancient*

History, ed. by R. S. Bagnall et al. (Oxford: Wiley-Blackwell, 2012) <https://doi.org/10.1002/9781444338386.wbeah18030>

FISCHER, IRENE, 'The Figure of the Earth — Changes in Concepts', *Geophysical Surveys*, 2 (1975), 3–54

FLINT, V. I. J., 'The "Elucidarius" of Honorius Augustodunensis and the Reform in Late Eleventh Century England', *Revue bénédictine*, 85 (1975), 178–89

——'Honorius Augustodunensis of Regensburg', in *Authors of the Middle Ages Historical and Religious Writers of the Latin West*, vol. 2: Nos. 5–6, ed. by C. J. Mews (Aldershot: Variorum, 1995), pp. 89–183

——'The Original Text of the "Elucidarius" of Honorius Augustodunensis from the Twelfth Century English Manuscripts', *Scriptorium*, 18 (1964), 91–94

——'The Sources of the "Elucidarius"', *Revue bénédictine*, 85 (1975), 190–98

——'World History in the Early Twelfth Century: The Imago Mundi of Honorius Augustodunensis', in *The Writing of History in the Middle Ages, Essays Presented to R. W. Southern*, ed. by R. H. C. Davis and J. M. Wallace-Hadrill (Oxford: Clarendon Press, 1981), pp. 211–38

FRANKLIN-BROWN, MARY, *Reading the World: Encyclopedic Writing in the Scholastic Age* (Chicago and London: University of Chicago Press, 2012)

FREEDMAN, PAUL, 'Spices and Late-Medieval European Ideas of Scarcity and Value', *Speculum*, 80 (2005), 1209–27

FRIEDMAN, JOHN BLOCK, *The Monstrous Races in Medieval Art and Thought* (Cambridge, MA: Harvard University Press, 1981)

FULTON, H., 'The Geography of Welsh Literary Production in Late Medieval Glamorgan', *Journal of Medieval History*, 41 (2015), 325–40

Geiriadur Prifysgol Cymru / A Dictionary of the Welsh Language (University of Wales, 2019), Online Version <http://www.geiriadur.ac.uk/gpc/gpc.html> [accessed 26 June 2020]

GERSH, S., 'Honorius Augustodunensis and Eriugena. Remarks on the Method and Content of the "Clavis Physicae"', in *Eriugena Redivius. Zur Wirkungsgeschichte seines Denkens im Mittelalter und im Übergang zur Neuzeit*, ed. by W. Beierwaltes (Heidelberg: Winter, 1987), pp. 162–73

GILL, CHRISTOPHER, 'Plato's Atlantis Story and the Birth of Fiction', *Philosophy and Literature*, 3 (1979), 64–78

GOETZ, HANS-WERNER, *Gott und die Welt. Religiöse Vorstellungen des frühen und hohen Mittelalters*, part I, volume 2 (Berlin: De Gruyter, 2012)

GOW, ANDREW, 'Gog and Magog on Mappaemundi and Early Printed World Maps: Orientalizing Ethnography in the Apocalyptic Tradition', *Journal of Early Modern History*, 2 (1998), 61–88

GRANT, EDWARD, *Planets, Stars, and Orbs: The Medieval Cosmos, 1200–1687* (Cambridge: Cambridge University Press, 1994)

GUY, BEN, 'A Welsh Manuscript in America: Library Company of Philadelphia, 8680.O', *National Library of Wales Journal*, 36 (2014), 98–123

HAHN, THOMAS, 'The Indian Tradition in Western Medieval Intellectual History', *Viator*, 9 (1978), 213–34

HAYCOCK, MARGED, ' "Some Talk of Alexander and Some of Hercules": Three Early Medieval Poems from the Book of Taliesin', *CMCS*, 13 (1987), 7–38

——'*Sy abl fodd, Sibli fain*: Sibyl in medieval Wales', in *Heroic Poets and Poetic*

Heroes in Celtic Tradition: A Festschrift for Patrick K. Ford, ed. by J. F. Nagy and L. E. Jones, CSANA Yearbook 3/4 (Dublin: Four Courts Press, 2005), pp. 115–30

HENG, GERALDINE, *The Invention of Race in the European Middle Ages* (Cambridge: Cambridge University Press, 2018)

Hereford Cathedral Mappa Mundi website <http://www.themappamundi.co.uk> [accessed 30 March 2016]

HIGGINS, IAIN MACLEOD, 'Defining the Earth's Center in a Medieval "Multi-Text": Jerusalem in the Book of John Mandeville', in *Text and Territory: Geographical Imagination in the European Middle Ages*, ed. by S. Tomasch and S. Gilles (Philadelphia: University of Pennsylvania Press, 1998), pp. 29–75

HOFMANN, J. B., *Lateinische Syntax und Stilistik*, 5th edn, rev. by A. Szantyr, 2 vols (Munich: C. H. Beck, 1965)

HOROBIN, SIMON, and JEREMY SMITH, *An Introduction to Middle English* (Edinburgh: Edinburgh University Press, 2002)

HUWS, DANIEL, 'Llyfr Coch Hergest', in *Cyfoeth y Testun. Ysgrifau ar Lenyddiaeth Gymraeg yr Oesedd Canol*, ed. by I. Daniel, M. Haycock, D. Johnston, and J. Rowland (Cardiff: University of Wales Press, 2003), pp. 1–30

—— *Medieval Welsh Manuscripts* (Aberystwyth: National Library of Wales, 2000)

IANNONE, A. PABLO, *Dictionary of World Philosophy* (London and New York: Routledge, 2001)

IMBACH, RUEDI, '*Empyreum*. Scholastische Gedanken über das Paradies', *Deutsches Dante-Jahrbuch*, 83 (2008), 13–37

JACKSON, KENNETH, 'The Date of the Old Welsh Accent Shift', *Studia Celtica*, 10/11 (1975/1976), 40–53

JONES, IDA B., 'Hafod 16', *ÉC*, 7 (1955), 46–75

—— 'Hafod 16 (A Medieval Welsh Medical Treatise) (suite et fin)', *ÉC*, 8 (1959), 346–93

JOOST-GAUGIER, CHRISTIANE L., *Measuring Heaven: Pythagoras and His Influence on Thought and Art in Antiquity and the Middle Ages* (Ithaca and London: Cornell University Press, 2006)

KASKE, R. E., ARTHUR GROOS, and MICHAEL W. TWOMEY, *Medieval Christian Literary Imagery: A Guide to Interpretation* (Toronto, Buffalo and London: University of Toronto Press, 1988)

KECK, DAVID, *Angels and Angelology in the Middle Ages* (Oxford: Oxford University Press, 1998)

KITCHELL, KENNETH F., JR, *Animals in the Ancient World from A to Z* (London and New York: Routledge, 2014)

KITSON, PETER R., 'Old English Literacy and the Provenance of Welsh *y*', in *Yr Hen Iaith: Studies in Early Welsh*, ed. by Paul Russell (Aberystwyth: Celtic Studies Publications, 2003), pp. 49–65

KLINE, NAOMI REED, *Maps of Medieval Thought: The Hereford Paradigm* (Woodbridge: Boydell Press, 2001)

LEFÈVRE, YVES, ed., *L'Elucidarium et les lucidaires* (Paris: E. de Boccard, 1954)

LE GOFF, JACQUES, 'Pourquoi le XIIIe siècle a-t-il été plus particulièrement un siècle d'encyclopédisme?', in *L'enciclopedismo medievale*, ed. by M. Picone (Ravenna: Longon, 1994), pp. 23–40

LEWIS, C. S., *The Discarded Image: An Introduction to Medieval and Renaissance Literature* (Cambridge: Cambridge University Press, 1994)

LEWIS, CHARLTON T., and CHARLES SHORT, *A Latin Dictionary* (Oxford: Clarendon Press, 1879)

LEWIS, HENRY 'Credo Athanasius Sant', *BBCS*, 5 (1930), 193–203

—— 'Ene, eny', *BBCS*, 1 (1921), 9–12

LIDDELL, HENRY GEORGE, and ROBERT SCOTT, *An Intermediate Greek-English Lexicon* (Oxford: Clarendon Press, 1945)

LLOYD-MORGAN, CERIDWEN, 'French Texts, Welsh Translators', in *The Medieval Translator 2*, ed. by Roger Ellis, Westfield Publications in Medieval Studies (London: Centre for Medieval Studies, Queen Mary and Westfield College, University of London, 1991), pp. 45–63

—— 'Rhai Agweddau ar Gyfieithu yng Nghymru yn yr Oesoedd Canol', *Ysgrifau Beirniadol*, 13 (1985), 134–45

LUCENTINI, PAOLO, *Platonismo medievale. Contributi per la storia dell'eriugenismo* (Florence: La nuova Italia, 1980)

LUFT, DIANA, 'Ansoddau'r trwnc: A Welsh Uroscopic Tract', *ZCP*, 58 (2011), 55–86

—— 'Awdur neu Dyallwr Ystoriâu: Theori a Chyfieithiadau Cymraeg yr Oesoedd Canol', *Llenyddiaeth mewn Theori*, 1 (2006), 15–40

—— 'Tracking ôl cyfieithu: Medieval Welsh Translation in Criticism and Scholarship', *Translations Studies*, 9 (2016), 168–82

MANZALAOUI, M. A., 'The *Secreta Secretorum* in English Thought and Literature from the Fourteenth Century to the Seventeenth Century with a Preliminary Survey of the Origins of the *Secreta*' (unpublished DPhil dissertation, University of Oxford, 1954)

MARTINDALE, CHARLES, *Redeeming the Text: Latin Poetry and the Hermeneutics of Reception* (Cambridge: Cambridge University Press, 1993)

MCKENNA, CATHERINE, 'Reading with Rhydderch: Mabinogion Texts in Manuscript Context', in *Language and Power in the Celtic World: Papers from the Seventh Australian Conference of Celtic Studies*, ed. by Anders Ahlqvist and Pamela O'Neill, Sydney Series in Celtic Studies 10 (Sydney: University of Sydney, 2011), pp. 205–30

MENEGALDO, SILVÈRE, 'Géographie et imaginaire insulaire au Moyen Âge, d'Isidore de Séville à Jean de Mandeville', *Les Lettres Romanes*, 66 (2012), 37–86

MERRILLS, ANDY, 'Geography and Memory in Isidore's *Etymologie*', in *Mapping Medieval Geographies: Geographical Encounters in the Latin West and Beyond, 300–1600*, ed. by Keith Lilley (Cambridge: Cambridge University Press, 2013), pp. 45–64

MEYER, H., 'Zum typologischen Grund der Triumphmetapher im „Speculum Ecclesiae" des Honorius Augustodunensis', in *Verbum et Signum. Beiträge zur mediävistischen Bedeutungsforschung*, ed. by H. Fromm, W. Harms, and U. Ruberg (München: Fink, 1975), pp. 45–58

MITTENDORF, INGO, 'Y Groglith Dyw Sul y Blodeu: Die mittelkymrische Passion nach Matthäus in Peniarth 14 und Havod 23', in *Übersetzung, Adaptation und Akkulturation im insularen Mittelalter*, ed. by Erich Poppe and Hildegard L. C. Tristram (Münster: Nodus Publikationen, 1999), pp. 259–88

MITTMAN, ASA SIMON, *Maps and Monsters in Medieval England* (New York and London: Routledge, 2006)

MITTMAN, ASA SIMON, and PETER DENDLE, eds, *The Ashgate Research Companion* to *Monsters and the Monstrous* (Farnham and Burlington: Ashgate, 2012)

MORGAN, T. J., 'Braslun o Gystrawen y Berfenw', *Bulletin of the Board of Celtic Studies*, 9 (1938), 195–215

MORRIS-JONES, JOHN, *Welsh Syntax: An Unfinished Draft* (Cardiff: University of Wales Press, 1931)

MUESSIG, CAROLYN, and AD PUTTER, eds, *Envisaging Heaven in the Middle Ages* (London and New York: Routledge, 2007)

MÜLLER, NICOLE, *Agents in Early Welsh and Early Irish* (Oxford: Oxford University Press, 1999)

MURDOCH, BRIAN, *The Medieval Popular Bible: Expansions of Genesis in the Middle Ages* (Cambridge: D. S. Brewer, 2003)

Ó GEALBHÁIN, SÉAMAS, 'The Double Article and Related Features of Genitive Syntax in Old Irish and Middle Welsh', *Celtica*, 22 (1991), 119–44

ORTOLEVA, VINCENZO, 'The Meaning and Etymology of the Adjective *Apiosus*', in *'Greek' and Roman in Latin Medical Texts*, ed. by Brigitte Maire, Studies in Cultural Change and Exchange in Ancient Medicine (Leiden: Brill, 2014), pp. 259–88

O'SULLIVAN, SINÉAD, 'The Corpus Martianus Capella: Continental Gloss Traditions on *De Nuptiis* in Wales and Anglo-Saxon England', *CMCS*, 62 (2011), 33–56

OWEN, MORFYDD E., 'Meddygon Myddfai: A Preliminary Survey of Some Medieval Medical Writing in Welsh', *Studia Celtica*, 10–11 (1975–1976), 210–33

—— 'The Medical Books of Medieval Wales and the Physicians of Myddfai', *The Carmarthenshire Antiquary*, 31 (1995), 34–44

PARINA, ELENA, 'The Polysemy of *llym* in Middle Welsh', *ZCP*, 63 (2016), 129–61

PARKER, W. H., 'Europe: How Far?', *The Geographical Journal*, 126 (1960), 278–97

PARKES, M. B., *Pause and Effect: An Introduction to the History of Punctuation in the West* (Aldershot: Ashgate, 1992)

PATTERSON, O., 'Honour and Shame in Medieval Welsh Society', *Studia Celtica*, 16–17 (1981–1982), 73–102

PÉPIN, JEAN, 'Recherches sur le sens et les origines de l'expression *Caelum Caeli* dans le Livre XII des *Confessions* de S. Augustin', *Archivum latinitatis medii aevi* (1953), 185–274

PETROFF, VALERY V., 'Eriugena on the Spiritual Body', *American Catholic Philosophical Quarterly*, 79 (2005), 557–610

PETROVSKAIA, N. I., 'Dating *Peredur*. New Light on Old Problems', *Proceedings of the Harvard Celtic Colloquium*, 29 (2009), 223–43

—— '*Delw y Byd*: la traduction médiévale en gallois d'une encyclopédie latine et la création d'un traité géographique', *ÉC*, 39 (2013), 257–77

—— 'La disparition du *quasi* dans les formules étymologiques des traductions galloises de l'*Imago Mundi*', in *La Formule au Moyen-Âge*, ed. by E. Louviot, ARTeM (Turnhout: Brepols, 2012), pp. 123–41

—— 'L'Image du monde animalier', in *Mondes animaliers au Moyen Age et à la Renaissance — Tierische Welten im Mittelalter und in der Renaissance — Actes du Colloque international des 8, 9, 10 et 11 mars 2016 à la Maison de la Culture d'Amiens*, ed. by Danielle Buschinger, Florent Gabaude, Marie-Geneviève Grossel, Jürgen Kühnel, and Mathieu Olivier (Amiens: Presses du Centre d'Etudes Médiévales de Picardie, 2016), pp. 313–26

——*Medieval Welsh Perceptions of the Orient* (Turnhout: Brepols, 2015)

—— 'Mythologizing the Conceptual Landscape: Religion and History in *Imago Mundi, Image du Monde* and *Delw y Byd*', in *Landscape and Myth in North-Western Europe*, ed. by M. Egeler (Turnhout: Brepols, 2019), pp. 195–211

—— '*Translatio* and *Translation*: The Duality of the Concept from the Middle Ages to the Early Modern Period', 同志社大学英語英文学研究 = *Doshisha Studies in English*, 99 (2018), 115–36 <https://doors.doshisha.ac.jp/duar/repository/ir/26025/020000990005.pdf >

—— 'The Travels of a Quire from the Twelfth Century to the Twenty-First: The Case of Rawlinson B 484, fols. 1–6', in *Middle English Texts in Transition: A Festschrift in Honour of Toshiyuki Takamiya*, ed. by M. Driver, L. Mooney, and S. Horobin (York: York Medieval Studies, 2014), pp. 250–67

PETROVSKAIA, N. I., and KIKI CALIS, 'Images of the World: Manuscripts Database of the Imago Mundi Tradition' <https://imagomundi.hum.uu.nl> (2019) [accessed 26 June 2020]

POPPE, ERICH, 'Exotic and Monstrous Races in the *Leabhar Breac*'s Gospel History and the Transmission of Arcane Knowledge to Medieval Ireland', in *Lochlann. Festskrift til Jan Erik Rekdal på 60-Årsdagen*, ed. by Cathinka Hambro and Lars Ivar Widerøe (Oslo: Hermes Academic Publishing, 2013), pp. 39–56

—— 'Owein, *Ystorya Bown*, and the Problem of "Relative Distance". Some Methodological Considerations and Speculations', *Arthurian Literature XXI: Celtic Arthurian Material*, ed. by Ceridwen Lloyd-Morgan (Woodbridge: D. S. Brewer, 2004), pp. 73–94

POPPE, ERICH, and R. RECK, 'A French Romance in Wales: *Ystorya Bown o Hamtwn*. Processes of Medieval Translations', *Zeitschrift für celtische Philologie*, 55 (2006), 122–80; 56 (2008), 129–64

—— 'Rewriting *Bevis* in Wales and Ireland', in *Sir Bevis of Hampton in Literary Tradition*, ed. by Jennifer Fellows and Ivana Djordjević (Cambridge: Boydell & Brewer, 2008), pp. 37–50

REISS, EDMUND, 'Welsh Versions of Geoffrey of Monmouth's *Historia*', *Welsh History Review*, 4 (1968), 97–127

REYNOLDS, ROGER E., 'Further Evidences for the Irish Origin of Honorius Augustodunensis', *Vivarium*, 8 (1969), 1–7

RIBÉMONT, BERNARD, *La «Renaissance» du XIIe siècle et l'Encyclopédisme*, Essais sur le Moyen Âge 27 (Paris: Champion, 2002)

ROBERTS, BRYNLEY F., 'The Treatment of Personal Names in the Early Welsh Versions of *Historia Regum Britanniae*', *BBCS*, 25 (1973), 274–89

—— 'Un o lawysgrifau Hopcyn ap Tomas o Ynys Dawy', *BBCS*, 22 (1967), 223–28

—— '*Ystoriaeu Brenhinedd Ynys Brydeyn*: A Fourteenth-Century Welsh Brut', in *Narrative in Celtic Tradition. Essays in Honor of Edgar M. Slotkin*, ed. by Joseph F. Eska, Proceedings of the Celtic Studies Association of North America, CSANA Yearbook 8–9 (Hamilton, NY: Colgate University Press, 2011), pp. 217–27

ROBERTS, RICHARD GLYN, 'Dalen Olaf Llawysgrif Peniarth 17', *Dwned*, 9 (2003), 37–42

ROCHE, JOHN, *The Mathematics of Measurement: A Critical History* (London and New York: Springer, 1998)

RODWAY, SIMON, 'Affectionate Cannibalism and the Blood Drinking Motif in Gaelic Literature', *CMCS*, 74 (2017), 47–65

—— 'Cymraeg vs. Kymraeg: Dylanwad Ffrangeg ar Orgraff Cymraeg Canol?', *Studia Celtica*, 43 (2009), 123–33

—— *Dating Medieval Welsh Literature: Evidence from the Verbal System* (Aberystwyth: CMCS, 2013)

—— 'The Red Book Text of "Culhwch ac Olwen": A Modernising Scribe at Work', *Studi Celtici*, 3 (2004), 93–161

ROWLAND, JENNY, 'The Manuscript Tradition of the Red Book *Englynion*', *Studia Celtica*, 18 (1983), 79–95

RUBIÉS, JOAN-PAU, *Travel and Ethnology in the Renaissance: South India through European Eyes, 1250–1625* (Cambridge: Cambridge University Press, 2000)

RUDDOCK, G., *Dafydd Nanmor* (Caernarfon: Gwasg Pantycelyn, 1992)

VAN RUITEN, J. T. A. G. M., 'The Four Rivers of Eden in the *Apocalypse of Paul* (*Visio Pauli*). The Intertextual Relationship of Genesis 2.10–14 and the *Apocalypse of Paul* 23', in *The Visio Pauli and the Gnostic Apocalypse of Paul*, ed. by Jan N. Bremmer and István Czachesz (Leuven: Peeters, 2007), pp. 50–76

RUSSELL, PAUL, 'The Joy of Six: Spelling and Letter-Forms among Fourteenth-Century Welsh Scribes' (forthcoming)

—— *Reading Ovid in Medieval Wales* (Columbus: Ohio State University Press, 2017)

—— 'Scribal (In)consistency in Thirteenth-Century South Wales: The Orthography of the Black Book of Carmarthen', *Studia Celtica*, 43 (2009), 135–74

RUSSO, L. C., 'The Reception of Medieval French Narrative in Medieval Wales: The Case of *Chwedyl Iarlles y Ffynnawn* and *Cân Rolant*' (unpublished PhD Thesis, University of Buenos Aires, 2014)

—— 'Translational Procedures in *Cân Rolant*, the Middle Welsh Translation of *La chanson de Roland*', *Brathair*, 14 (2014), 109–28

SAENGER, PAUL, *Space Between Words: The Origins of Silent Reading* (Stanford: Stanford University Press, 1997)

SANFORD, EVA MATTHEWS, 'Honorius, *Presbyter* and *Scholasticus*', *Speculum*, 23 (1948), 397–425

SCHERB, VICTOR I., 'Assimilating Giants: The Appropriation of Gog and Magog in Medieval and Early Modern England', *Journal of Medieval and Early Modern Studies*, 32 (2002), 59–84

SCHUMACHER, S., 'Mittel- und Frühneukymrisch', in *Brythonic Celtic — Britannisches Keltisch: From Medieval British to Modern Breton*, ed. by Elmar Ternes, Münchner Forschungen zur historischen Sprachwissenschaft, Bd. 11 (Bremen: Hempen, 2011), pp. 85–236

VAN SEVENTER, NELY, 'The Translation of the *Sybilla Tiburtina* into Middle Welsh', in *'Y Geissaw Chwedleu': Proceedings of the 7th International Colloquium of Societas Celto-Slavica*, ed. by Aled Llion Jones and Maxim Fomin (Bangor: Bangor University, 2018), pp. 109–18

SHAPIRO, M., 'Perseus and Bellerophon in "Orlando Furioso"', *Modern Philology*, 81 (1983), 109–30

SIMEK, RUDOLF, *Heaven and Earth in the Middle Ages: The Physical World Before Columbus*, trans. by Angela Hall (Woodbridge: Boydell, 1996)

SIMS-WILLIAMS, PATRICK, 'Variation in Middle Welsh Conjugated Prepositions: Chronology, Register and Dialect', *Transactions of the Philological Society*, 111 (2013), 1–50

SMALLEY, STEPHEN S., 'Spiritual Gifts and I Corinthians 12–16', *Journal of Biblical Literature*, 7 (1968), 427–33

SOUTHERN, R. W., *Saint Anselm: A Portrait in a Landscape* (Cambridge: Cambridge University Press, 1990)

—— *St Anselm and His Biographer* (Cambridge: Cambridge University Press, 1963)

SPRINBORG, P., 'Weltbild mit Löwe. Die *Imago mundi* von Honorius Augustodunensis in der Altwestnordischen Textüberlieferung', in *Cultura Classica e Cultura Germanica Settentrionale. Atti del Convegno Internazionale di Studi Università di Macerata, Facoltà di Lettere e Filosofia, 2–4 maggio 1985*, ed. by Pietro Janni, Diego Poli, and Carlo Santini (Rome: Herder, 1988), pp. 167–219

STEADMAN, JOHN M., 'Perseus upon Pegasus' and Ovid Moralized', *The Review of English Studies*, 9 (1958), 407–10

STEVENSON, JANE, 'Altus Prosator', *Celtica*, 23 (1999), 326–68

STUMP, E., ed., *The Cambridge Companion to Augustine* (Cambridge: Cambridge University Press, 2001)

STURLESE, LORIS, 'Zwischen Anselm und Johannes Scotus Eriugena: der seltsame Fall des Honorius, des Mönchs von Regensburg', in *Historia Philosophiae Medii Aevi*, ed. by Burkhard Mojsisch and Olaf Pluta, 2 vols (Amsterdam: John Benjamins, 1991), II, pp. 927–51

SVANHILDUR ÓSKARSDÓTTIR, 'Prose of Christian Instruction', in *A Companion to Old Norse-Icelandic Literature and Culture*, ed. by Rory McTurk, Blackwell Companions to Literature and Culture 31 (Malden MA, Oxford and Victoria: Blackwell, 2005), pp. 338–53

TAMANAHA, BRIAN Z., *On the Rule of Law: History, Politics, Theory* (Cambridge: Cambridge University Press, 2004)

TATTERSALL, JILL, 'Anthropophagi and Eaters of Raw Flesh in French Literature of the Crusade Period: Myth, Tradition and Reality', *Medium Aevum*, 57 (1988), 240–53

THOMAS, PETER WYNN, 'Cydberthynas y pedair fersiwn ganoloesol', in *Canhwyll Marchogyon: Cyd-destunoli 'Peredur'*, ed. by Sioned Davies and P. W. Thomas (Cardiff: University of Wales Press, 2000), pp. 10–49

—— *Gramadeg y Gymraeg* (Cardiff: University of Wales Press, 1996)

—— 'Middle Welsh Dialects: Problems and Perspectives', *BBCS*, 40 (1993), 17–50

—— '(-th-): Tystiolaeth Beirdd y Tywysogion a'r Uchelwyr', *Dwned*, 15 (2009), 11–32

TRISTRAM, HILDEGARD L. C., 'Der *homo octipartitus* in der irischen und altenglischen Literatur', *ZCP*, 34 (1975), 119–53

TRY, REBECCA, 'NLW MS 5267B; A Partial Transcription and Commentary' (unpublished MPhil dissertation, Cardiff University, 2015)

UNDERDOWN, RUTH CARYS, 'Studies in Welsh Prepositions: After' (unpublished PhD thesis, University of Cambridge, 2008)

VAN DUZER, CHET, '*Hic sunt dracones*: The Geography and Cartography of Monsters', in *The Ashgate Research Companion to Monsters and the Monstrous*, ed. by Asa Simon Mittman and Peter J. Dendle (Farnham: Ashgate, 2012), pp. 387–435

WAGNER, RUDOLF G., 'Does This Make Sense? Reading Sinological Translations', in *Zurück zur Freude. Studien zur chinesischen Literatur und Lebenswelt und ihrer Rezeption in Ost und West. Festschrift für Wolfgang Kubin*, ed. by Marc

Hermann, Christian Schwermann, and Jari Grosse-Ruyken (Sankt Augustin: Institut Monumenta Serica, 2007), pp. 767–76

WILLIAMS, GRUFFYDD ALED, 'Mwy am Lawysgrif Gymraeg yn U.D.A.: The Public Library Company of Philadelphia, Llsgr. 8680.O', *Llên Cymru*, 34 (2011), 248–50

WILLIAMS, MARK, *Fiery Shapes: Celestial Portents and Astrology in Ireland and Wales, 700–1700* (Oxford: Oxford University Press)

WILLIAMS, STEPHEN J., *The Secret of Secrets: The Scholarly Career of a Pseudo-Aristotelian Text in the Latin Middle Ages* (Ann Arbor: University of Michigan Press, 2003)

WILLIS, DAVID W. E., *Syntactic Change in Welsh* (Oxford: Clarendon Press, 1998)

WISNOVSKY, ROBERT, FAITH WALLIS, JAMIE C. FUMO, and CARLOS FRAENKEL, eds, *Vehicles of Transmission, Translation, and Transformation in Medieval Textual Culture*, Cursor Mundi 4 (Turnhout: Brepols, 2011)

WITTKOWER, RUDOLF, 'Marvels of the East. A Study in the History of Monsters', *Journal of the Warburg and Courtauld Institutes*, 5 (1942), 159–97

WOLF, KIRSTEN, 'The Colors of the Rainbow in Snorri's *Edda*', *Maal og Minne*, 1 (2007), 51–62

WOODWARD, DAVID, 'Medieval *Mappaemundi*', in *History of Cartography I: Cartography in Prehistoric, Ancient, and Medieval Europe and the Mediterranean*, ed. by J. B. Harley and David Woodward (Chicago and London: University of Chicago Press, 1987), pp. 286–370

—— 'Reality, Symbolism, Time, and Space in Medieval World Maps', *Annals of the Association of American Geographers*, 75 (1985), 510–21

YINGST, DANIEL, '*Quae Omnia Concorditer Consonant*: Eriugena's Universe in the Thought of Honorius Augustodunensis', in *Eriugena and Creation: Proceedings of the Eleventh International Conference on Eriugenian Studies, held in honour of Edouard Jeauneau, Chicago, 9–12 November 2011*, ed. by W. Otten and Michael I. Allen, Instrumenta Patristica et Mediaevalia 68 (Turnhout: Brepols, 2014), pp. 427–61

ZUCKER, ARNAUD, 'Introduction', in *Encyclopédire: Formes de L'Ambition Encyclopédique dans l'Antiquité et au Moyen Âge*, ed. by Arnaud Zucker, Collection d'Études Médiévales de Nice 14 (Turnhout: Brepols, 2013), pp. 11–28